SpringerBriefs in Ethics

Springer Briefs in Ethics envisions a series of short publications in areas such as business ethics, bioethics, science and engineering ethics, food and agricultural ethics, environmental ethics, human rights and the like. The intention is to present concise summaries of cutting-edge research and practical applications across a wide spectrum.

Springer Briefs in Ethics are seen as complementing monographs and journal articles with compact volumes of 50 to 125 pages, covering a wide range of content from professional to academic. Typical topics might include:

- Timely reports on state-of-the art analytical techniques
- A bridge between new research results, as published in journal articles, and a contextual literature review
- A snapshot of a hot or emerging topic
- In-depth case studies or clinical examples
- Presentations of core concepts that students must understand in order to make independent contributions

Anne Zimmerman

Informed Consent for AI and Health Care

Consumer Rights and Industry Obligations

Springer

Anne Zimmerman
School of Professional Studies
Columbia University
New York, NY, USA

ISSN 2211-8101 ISSN 2211-811X (electronic)
SpringerBriefs in Ethics
ISBN 978-3-032-11346-7 ISBN 978-3-032-11347-4 (eBook)
https://doi.org/10.1007/978-3-032-11347-4

This Springer imprint is published by the registered company Springer Nature Switzerland AG
The registered company address is: Gewerbestrasse 11, 6330 Cham, Switzerland

Preface

There is some fear around the swift pace of adoption of AI in the healthcare industry. Some authors (bioethicists, lawyers, clinicians, etc.) are quick to point out the risks. Others tout AI as a panacea. AI has lots to offer healthcare consumers. In writing this, I approached informed consent from the question of what best serves the healthcare consumer, thinking a lot about the interests of those in need of medical treatments. To me, informed consent is not a barrier to innovation or to treatment. While it may carry some inconvenience, it is a way to forge a stronger partnership between clinicians and healthcare consumers.

New York, USA

Anne Zimmerman

About This Book

This book identifies areas of uncertainty about informed consent, and specifically, the disclosures necessary for the ethical use of AI in health care. The level of detail necessary for obtaining informed consent varies according to the AI tool in question. The framework provided here should foster consideration of risks that AI tools pose and the likelihood of a clinician overly relying on AI. The growing use of AI in health care for diagnosis, medical notes, drug discovery, and more has led to open questions about disclosure to the public as well as to healthcare consumers. There is no legally binding universal right to know whether or how AI is used in health care. The traditional right to informed consent requires disclosure, often based on what a consumer would should know prior to making healthcare decisions, for example, the likelihood of debilitating side effects. The system of informed consent as disclosure, understanding, and voluntary agreement as commonly used in health care differs from informed consent to release one's data, governed by a web of privacy laws and institutional policies, also impacted by the growing use of AI. And furthermore, disclosure of the role of AI in drug discovery or in medical records brings different concerns for which the traditional person-centered informed consent concept may appear insufficient or inapplicable. However, informed consent is adaptable and is often the best approach to AI in health care. Transparency may be ethically and legally required even when the healthcare consumer's sign-off is not required. This book explores scenarios in which consumers and the public should have some level of information about the risks, benefits, and inner workings of AI systems used in the course of medical care, and aims to explain why, arguing that informed consent is appropriate and beneficial for many uses of AI.

Note on Language

"Healthcare consumer" is used more than the word patient as the word patient connotes someone deferential or passive. Nonetheless, especially when referring to cases or journal articles, patient is used occasionally to align the word use with that of other authors. "Clinician" is used as a broad catchall for physicians, nurse practitioners, physician's assistants, nurses, scientists, and technicians who perform many duties, including duties that are behind the scenes and not directly facing the consumer, some of which were reserved for physicians in the past. As a stylistic choice, "data" is treated as singular rather than plural in accordance with modern journalistic guidelines. "Healthcare" is used as an adjective and "health care" as a noun with descriptor throughout.

Contents

About the Author

Anne Zimmerman, JD, MS, CIPP-US (Certified Information Privacy Professional), CEET (Certified Ethical Emerging Technologist) is a Lecturer in AI & Ethics, Columbia University; Editor-in-Chief of *Voices in Bioethics* (Columbia University journal and podcast); Advisory Board Member, Columbia University Master of Science in Bioethics Program; Chair of the New York City Bar Association Bioethical Issues Committee; Co-Chair, AI and Healthcare Subcommittee of the New York City Bar Association Presidential Task Force on Artificial Intelligence and Digital Technologies; Editor, Digital Society Topical Collection "Exploring technologies as a social determinant of health;" Co-Founder and Chair of *Innovative Bioethics Forum*; Consultant to MyBioethics (ed-tech app); a Spring 2023 Colorado College Innovator in Residence; and a mentor for All Tech Is Human. Her book, *Medicalization: An Encroachment on Consent, Culture*, examines the phenomenon of medicalization and the increasingly large, invasive, and coercive role of medicine in society. Her previous book, *Medicine, Power, and the Law: Exploring a Pipeline to Injustice* explores the relationships between medicine, science, and technology and the criminal and civil justice systems.

Chapter 1
Understanding AI for Health Care

1.1 Introduction

Most people seeking medical care want to know basic information about the care they will receive. For example, if you are having a hip replacement, you may want to know how long the recovery time will be, when you can exercise again, and whether an infection or lifelong limitation could ensue. The law requires clinicians to disclose risks and benefits. That disclosure allows people seeking or refusing care to make informed decisions when they consent to or refuse medical treatments. However, there is no current prevailing framework for identifying when informed consent prior to a treatment must include disclosures about whether and how AI was used. Clinicians and researchers use AI in a variety of ways including drug discovery, clinical trials, diagnosis, prognosis, treatment planning, surgery, triage, hospital admissions and discharges, and administration. Disclosures for the purpose of informed consent in health care are different from disclosures required by the principle of transparency commonly applied in AI ethics. Many disclosures to the public or disclosures to consumers (even of health care) do not require the consumer to provide explicit consent, or they require merely a cursory sign-off, like checking a box. Some consents are opt-out rather than opt-in and are quite lax. When AI is used in healthcare settings, clear disclosure requirements would resolve some uncertainty and protect consumers of health care. Informed consent's relevance depends on characteristics of the AI tool itself, how it is used, whether a reasonable person seeking health care would find knowledge about its use relevant to decision making (materiality), and the degree and type of potential harm. When specific individualized informed consent is not possible, legally required, or ethically compelled, other transparency-based safeguards should protect and inform consumers and the public.

This book aims to resolve some of the uncertainty about disclosing the use of AI by offering a framework for deciding upon the level of detail necessary in disclosures. There is no declared and legally binding universal right to know how AI is used

A. Zimmerman, *Informed Consent for AI and Health Care*,
SpringerBriefs in Ethics, https://doi.org/10.1007/978-3-032-11347-4_1

in health care. Yet to ethically employ AI in many healthcare situations, informed consent should be standard practice.

Disclosure is key to the traditional right to informed consent. Courts often base the need for disclosure on what a reasonable person would want or expect to know prior to making healthcare decisions. Yet some jurisdictions continue to base disclosure requirements on what a reasonable physician would disclose, i.e., a community standard. The system of informed consent as disclosure, understanding, and agreement as commonly used in health care differs somewhat from informed consent to release one's data, governed by a web of consumer privacy laws and institutional policies. The data landscape has changed with the growing use of AI and the massive datasets used to train AI tools. And furthermore, disclosure of the role of AI in pre-clinical-trial drug discovery or in medical records brings different concerns for which the traditional patient-centered informed consent concept may at first appear insufficient or inapplicable. However, informed consent is adaptable and is the best approach to AI in health care.

This book explores scenarios in which those consuming health care and/or the public ought to know how AI is used in their health care and aims to explain why, arguing that informed consent is a healthcare-consumer-friendly approach to the use of AI in health care. Then, it provides a simple framework advocating for levels of disclosure based on risk and the likelihood that clinicians will overly rely on AI.

1.2 AI Developments

People often assume that AI is new and requires all new frameworks. The term AI was coined in the 1950s when computer programmers first succeeded in designing a machine to perform rudimentary logic. AI simulates human intelligence processes using machines or computer systems. It can mimic some human abilities and perform problem-solving, predictive, and creative tasks.[1] Yet AI operates through mathematical instructions and is unlike human thought, cognition, consciousness, or awareness.[2] Repeated mathematical operations lead to an output. AI sifts through data using complex layers of operations to make predictions, and more recently, to generate text, images, and other new content.

Since the Logic Theorist program was created in the mid-1950s,[3] AI has changed significantly. The defining characteristics of modern AI are large datasets and

---

[1] Landgrebe, J. and Smith, B. (2022). *Why Machines Will Never Rule the World*. Routledge; also a note on the word creative—like the word think or understand, we do not have a vocabulary for actions AI tools take on that had been reserved for humans. For AI, creative does not imply the inspired way that artists create, it means produces an output that is in the category of creative works, like reports, books, paintings, sketches, poetry, etc.

[2] Landgrebe and Smith, 2022.

[3] Sloat, S. (2023) The first AI started a 70-year debate: The Logic Theorist started a discussion that continues today—can a machine be intelligent like us? Popular Science. https://www.popsci.com/technology/the-first-ai-logic-theorist/.

complex algorithms. AI includes many types of systems and is not limited to its newer iterations like generative AI, large language models, or natural language processing. AI has been used in medicine for years. Throughout the 1980s and 90s, AI was used to manage data and records, and in surgery and research. As AI became more complex, advanced, and available, its use in radiology and in electronic health records systems (EHR) grew considerably. With the advent of big data and increasingly complex AI, healthcare systems continue to increase the use of AI, especially using new AI tools in genetics, bioinformatics, precision medicine, complex surgeries, public health, and clinical decision support. With each new purchase or upgrade to existing technology, hospitals rely more and more on AI, often without disclosure or consent. AI's arrival may be somewhat stealth due to upgrades—it is not necessarily scrutinized for ethical concerns prior to purchase or deployment. The procurement department sometimes has as much impact on the decision to purchase an AI tool as medical teams do. Furthermore, salespeople have various ways to market products, often through sponsorship at conferences of large medical associations. There is normally not an ethicist at the purchasing point.

The changes to AI bring additional ethical issues. Because older tools may have been rudimentary, they did not have as large an impact on consumer outcomes as modern AI tools are likely to have. Big data leads to ethical issues that were almost irrelevant to localized data collection and storage. For example, now tech companies use nearly all data freely found on the internet to train generative AI tools. Widely deployed AI tools that map genomes, classify images, or direct public health policy can impact more people more quickly than older AI tools. This high impact is coupled with less explainability. Therefore, ethical deployment requires some balance between caution and charging ahead.

1.3 AI Inner Workings

Understanding AI is crucial to being able to explain it to healthcare consumers. The very basics provide a good starting point. AI models use data and algorithms to provide output. Within AI, machine learning is an important subcategory. Machine learning classifies, categorizes, groups, predicts, suggests, recommends, and provides possible outcomes. Machine learning uses algorithms to identify patterns in data and use them to create a model that can make predictions. Supervised learning models are trained on labeled data—they sort and classify. Unsupervised learning clusters unlabeled data, for example, unlabeled images. In machine learning, mathematical models adjust without direct instruction, allowing their accuracy to improve over time with experience and added data.

1.3.1 Data

Data is a specific type of information. It has broad and narrow meanings. Some use data in a medical context to suggest proof—"just look at the data"—while in the AI context, data means individual pieces of information. Data provides the basis (the ingredients) for computational reasoning, calculations, operations, and AI algorithms.

Data can be raw, meaning it is not arranged or organized by class, genre, or type, etc. Data can have an apples and oranges feel. It includes numbers, words, text, images, video, measurements, symbols, and other types of raw information. Data comes from many sources including consumers of healthcare products and services. Sometimes a human enters data; other times data is captured by devices including ambient systems and wearables. Lots of data exists on the internet, which is to some degree being repurposed to serve the needs of AI.[4] Common crawl and colossal clean crawled corpus (C4) are examples of data scraped from the internet. There are active debates about intellectual property rights infractions and this repurposing of the internet to capture its data. Restrictions on data for training AI seem to be increasing.[5]

Additionally, entities that collect data (for example, hospital systems, consumer corporations (online and brick and mortar), nonprofits, etc.) sell data. Once data is entered or drawn into any dataset, it may be sold or combined with data from other sources, sometimes illegally and other times as part of the 260-billion-dollar market for data. Data brokers aggregate, organize or process, license, and sell data.

Data is crucial to understanding AI ethics. Some data is personal, private, or incriminating. Medical providers must be proper stewards of personal data. They have a duty to protect privacy. There is a lifespan of data—policies should cover data collection, transfer and storage, encryption, all uses, and disposal. Data must make sense over time or be disposed of or stored in organized forms with timestamps or dates.

For the purposes of AI, especially that used in health care, large datasets are the basis for recommendations. Therefore, data must be accurate. For example, if a person is six feet tall, the "6" may be entered as a point of data. If the person is really 5 feet, the AI system would be relying on false data when it recommends a drug dosage.

Some data is not reliable, reflects noise, includes outliers, was entered incorrectly, reflects bias of the person who entered it, or is otherwise corrupt. Data also may be the documentation of judgment or opinion. Distinguishing opinion from fact is a crucial aspect of critical thinking. Yet it is often ignored. For example, data may be numbers based on the pain scale, a common tool by which a healthcare consumer describes his or her own pain on a scale of one to ten. If a dataset has 100,000

---

[4] Longpre, S., Mahari, R., et al. (2024). Consent in Crisis: The Rapid Decline of the AI Data Commons. *Data Provenance Initiative* (MIT). https://www.dataprovenance.org/consent-in-crisis-paper.

[5] Longpre, et al., 2024.

people's pain scale data, what does it really say? The pain scale is subjective—there is no way to determine what an eight means to one person compared to what it means to another. If the pain opinions were ascertained post-surgery, there may be confounding variables that are left out. Often, numbers begin to look like fact. There may be a distribution of pain scale opinions—it may look like a normal distribution (bell shaped); it may skew to high pain or low pain. A different set of numbers may include blood pressure or observable behaviors and actions which can act as a proxy for pain. A statistician or clinician trying to create a pain regimen may overvalue the AI despite its limitations as pain, moral values concerning drugs for pain, and resistance to or efficacy of drugs vary considerably.

Data, information, knowledge, wisdom (DIKW) is a framework for looking at how data becomes meaningful. It can be seen as a theory for understanding data use and how proper inferences can be drawn from data. DIKW is usually presented as a pyramid and has been used across industries and the military.**[6] Raw data is not always useful in a vacuum, but human or AI manipulation of data carries risks. Contextualization is another important concept for the evaluation of data. Increasingly sophisticated AI can address context to a degree with access to context-relevant data. Engineers aim to input context making AI tools industry specific, more meaningful to users, and more productive.

1.3.2 Algorithms

Algorithms are like instructions. They are a sequence of rules followed to create an output. The algorithm is mathematical and can take many forms to perform tasks (for example, regression, classification, clustering). The algorithm governs the process—data is the input—the output depends on the type and purpose of the model. There are many ways to express an algorithm to help users understand the AI model, for example, a flowchart. Algorithms are expressed in code but with machine learning adapt over time and/or based on new data. Programming languages like Python allow engineers to create sets of instructions for the model. (Pseudocode is a technique to allow those without significant coding or engineering experience to understand the steps of an algorithm. It is also used to sketch out an algorithm.)

Due to the complexity of AI models, the same algorithm may lead to different output despite the same inputs. As information passes through computational layers, for example, in neural networks, models can adapt and draw inferences from the data without specific instructions. Machine learning means the algorithm self-adjusts, and models may change over time as new data is provided. Therefore, the weight of each factor may change over time through machine learning as well as through engineers

[6] Rowley, Jennifer (2007). "The Wisdom Hierarchy: Representations of the DIKW Hierarchy". *Journal of Information and Communication Science.* 33 (2): 163–180. https://doi.org/10.1177/0165551506070706. S2CID 17000089.

tweaking the model to improve it. This complexity is highly relevant to the use of AI in health care.

1.3.3 Large Language Models and Generative AI

Generative AI is so named because it is a type of AI that generates new content based on training data. Like other AI, it uses data to predict and then, unlike other machine learning models, it produces the new content. The generated content may be a book, video, image, synthetic cell samples, or even code. For example, it creates artwork, stories, replicas, meeting note summaries, research summaries, and health-care consumer personas. ChatGPT and other generative AI terminology can cause confusion. (ChatGPT is a generative AI tool, yet it has some nongenerative uses. ChatGPT can retrieve information, translate, or fact-check based on a dataset, but it is designed mostly to create. In discussing machine learning, I avoid the phrase the AI *generated* a prediction or recommendation as it causes confusion.)

Natural language processing, which allowed human language like free form text and speech to be processed, organized, and used as data for machine learning, greatly improved the capability of computer systems. Machine learning and natural language processing laid the ground for generative AI, which uses considerably more power and sophisticated technology. Large language models (LLMs) are deep learning AI models.[7] They perform tasks and create new content (generative AI). (They are just one category of generative AI.)[8] These models appear to "understand" natural language, but they are not capable of understanding in the same way in which humans understand. They operate on patterns and prediction, mimicking but not engaging in human reasoning.[9] Generative AI essentially predicts what comes next in a sequence based on vast training data. But it does so through convoluted means. It uses mathematical representations. New research suggests LLMs have complex ways of arriving at even simple answers. For example, basic mathematical tasks go through many steps. And when completing or creating sentences, LLMs may approach the task from many angles.[10]

LLMs are generally used to generate creative output. Rather than being exclusively logic-based, LLMs scour the data for word associations. This means an LLM AI tool is unlikely to choose the exact same response twice. If one were to use it to write a

---

7 Glossary of Terms: Generative AI Basics. MIT Management. https://mitsloanedtech.mit.edu/ai/basics/glossary/#Large_Language_Model_LLM.

8 Sandhu, J. (2024). What are LLMs and generative AI? Schwartz Reisman Institute for Technology and Society. University of Toronto. https://srinstitute.utoronto.ca/news/gen-ai-llms-explainer.

9 Landgrebe and Smith, 2022.

10 Laird, J. (2025). Anthropic has developed an AI 'brain scanner' to understand how LLMs work and it turns out the reason why chatbots are terrible at simple math and hallucinate is weirder than you thought. PC Gamer. https://www.pcgamer.com/software/ai/anthropic-has-developed-an-ai-brain-scanner-to-understand-how-llms-work-and-it-turns-out-the-reason-why-chatbots-are-terrible-at-simple-math-and-hallucinate-is-weirder-than-you-thought/.

paper or a book, it would be nearly impossible to generate the same response again given the same prompt. Generative AI is predictive—predicting the next word, the next image, the next sequence, based on big data. If training data includes the phrase "Sudoku is difficult" ten million times, and someone prompts "Sudoku is" it is likely to respond "difficult." With big data, the range of possible creations is very wide.

Generative AI includes LLMs as well as generative adversarial networks (GANs) composed of two neural networks, a generator and a discriminator. The generator analyzes the training data for attributes. The discriminator analyzes the training data and distinguishes fake from real, or predicts whether an image belongs in the group, providing feedback to the generator.[11] GANs are used to generate images and videos and are increasingly important in medical research.[12]

1.4 AI Basics for Bioethicists/non-Techies

While health care entities have robust information technology (IT) departments, those using AI in some clinical settings may want to understand the basic inner workings and architecture of the tools they use. For details on AI, several text books provide background information and concepts as well as computational and mathematical details.[13] Some rules of thumb are helpful. Humans generally have a good sense of what the purpose of the AI is and what type of output it will deliver. Clinicians know whether they are using AI to match a genetic mutation to a drug or to classify a cancer based on MRI scans. And the most common uses are to relieve an administrative burden. These three are a shortcut for a minimum understanding of AI:

- first, the AI model and architecture act as a roadmap,
- second, data is like fuel, and,
- third, the algorithm is the instruction manual, with the important twist being that it may be dynamic.

Special dichotomies help clinicians and bioethicists understand AI basics. There is a crucial distinction between generative AI and other machine learning tools. AI tools that classify an image differ from those that produce an image. Many AI tools in health care use machine learning to predict, categorize, classify, or recommend. These tools are not generative. Decision support systems and software used as a medical device often use machine learning tools that are not large language models. Much of the diagnostic and prognostic AI in medicine is machine learning and not

[11] What is a GAN? Amazon Web Services. https://aws.amazon.com/what-is/gan/ "The discriminator calculates the probability that the generated output belongs to the original dataset".

[12] Hansen, C. (2022). Generative adversarial networks explained. IBM. https://developer.ibm.com/articles/generative-adversarial-networks-explained/.

[13] Russell, S. J., & Norvig, P. (2020). *Artificial Intelligence: A Modern Approach* (4th ed.). Pearson; Hamadani, A., Ganai, N., Henna, H., Bashir, J. (2024). *A Biologist's Guide to Artificial Intelligence: Building the foundations of Artificial Intelligence and Machine Learning for Achieving Advancements in Life Sciences*. Academic Press.

generative. Physicians creating meeting notes may rely on generative AI tools. For example, a machine learning tool may recommend five options to a clinician who then may apply a generative AI tool to communicate with the healthcare consumer.

There is a different distinction between fully explainable systems using local or accessible datasets and complex machine learning systems based on large datasets (big data). In medicine, many AI products (software as a medical device and clinical decision support tools) that are not completely and easily explainable and use big data pose ethical issues. There are also tools designed to help explain otherwise unexplainable AI models.[14]

Additionally, the broad term "AI" includes any logic-based computer program. A hospital collecting its own data and comparing current healthcare consumers to past ones treated for the same illness in the same facility may create an AI tool to make recommendations or diagnoses. The programmed software systems may be entirely explainable and not use highly sophisticated machine learning. Most importantly, some AI is programmed and explainable; some AI is not explainable. And some tools fall in between, are partially explainable, or can be explained using technical tools designed to crack the black box.

As a quick analytical shortcut, these ways of categorizing AI tools have proven helpful:

- generative/nongenerative (tools that predict, recommend, classify, categorize),
- local dataset/big data,
- machine learning/programmed,
- black box/explainable.

While these oversimplify AI, especially considering complex tools, they nonetheless prove helpful to those looking to organize their thoughts and analysis of AI. Understanding the basics of AI is crucial to the disclosure necessary to obtain informed consent from healthcare consumers. The basics help facilitate communication about AI. Incorrect terminology like using "AI" to mean only generative AI, (for example, "AI is brand new technology used to write papers or create images,") or using "generative AI" to mean all machine learning AI (for example, suggesting that all clinical decision support tools are generative AI or overusing the term ChatGPT) can cause misunderstandings. To be transparent about AI and to make the disclosures appropriate and necessary for informed consent, clinicians will need to use correct terminology and be transparent about their own knowledge and limitations.

[14] Ali, S., Abuhmed, T., et al. (2023). Explainable Artificial Intelligence (XAI): What we know and what is left to attain Trustworthy Artificial Intelligence. *Information Fusion.* Volume 99, November 2023. https://www.sciencedirect.com/science/article/pii/S1566253523001148?via%3Dihub.

Chapter 2
Transparency

2.1 Transparency Differs from Informed Consent

Transparency is one-directional as a principle, although as applied, consent in the form of a click or sign-off may be required to acknowledge a disclosure. Informed consent in medicine is always two-directional—the recipient of the information consents to or rejects the surgical procedure, drug, or other medical intervention. *Informed consent is a handshake while transparency is an announcement*. This will be important for determining the interplay between transparency and informed consent. Transparency is an AI principle to protect the public and consumers, including health-care consumers. Yet transparency is not enough to ensure that individuals' rights are protected during the course of medical treatment or clinical research.

2.2 Transparency

Transparency is an AI ethics buzzword, part of nearly all sets of AI principles.[1] The US Office of Science and Technology Policy's Blueprint for an AI Bill of Rights included a right to notice and explanation:

> You should know that an automated system is being used and understand how and why it contributes to outcomes that impact you.[2]

---

[1] Fjeld, J., Achten, N., Hilligoss, H., et al. (2020) Principled Artificial Intelligence: Mapping Consensus in Ethical and Rights-based Approaches to Principles for AI. *The Berkman Klein Center for Internet & Society Research Publication Series.* Research Publication No. 2020–1 https://cyber.harvard.edu/publication/2020/principled-ai.

[2] Office of Science and Technology. AI Bill of Rights. White House Archives. https://bidenwhitehouse.archives.gov/ostp/ai-bill-of-rights/#notice (a Biden executive order based on the AI Bill of Rights was enacted and later rescinded).

A. Zimmerman, *Informed Consent for AI and Health Care*,
SpringerBriefs in Ethics, https://doi.org/10.1007/978-3-032-11347-4_2

The Bill of Rights is consumer and public focused. It imposes an industry obligation to provide straightforward "descriptions of the overall system functioning and the role automation plays, notice that such systems are in use, the individual or organization responsible for the system, and explanations of outcomes that are clear, timely, and accessible."[3]

Transparency is less precise than informed consent. It is a sometimes-limited principle with its own frameworks, open to many interpretations. It can seem wishy washy in contrast to informed consent, which is an ethical and legal requirement in health care. Transparency is both a part of informed consent (disclosure) and a justification for it (a normative concept holds that transparency is preferable to secrecy or nondisclosure). Transparency is a principle, an idea, and an aspiration. In many contexts, it is a legal requirement.[4] For example, the EU's General Data Protection Regulation and the California Consumer Privacy Act require it. The Centers for Medicare and Medicaid Services (CMS) and professional associations including the American Medical Association do as well. However, the steps required to be transparent are often debated. Transparency is not entirely objective, especially for circumstances in which it is not required by law, but soft law, nonbinding guidelines, and institutional practices call for it. There are differences of opinion about the amount of detail one must disclose to label oneself transparent.

The principle of transparency in AI imposes a duty on humans. The duty of transparency requires disclosures, including making those impacted by AI aware of its use. The duty does not require a meeting of the minds or documentation to evidence consumer understanding of the information that was disclosed. In many online situations, transparency about data use, cookies, or tracking pixels may require a click as a cursory sign-off. That click often feels like a quid pro quo as it allows consumers access to online materials and websites, etc. Rejecting some cookies is now easier than ever due to legal requirements distinguishing required cookies from those used for advertising and other purposes.

According to the Organisation for Economic Co-operation and Development (OECD),

> *AI Actors should commit to transparency and responsible disclosure regarding AI systems. To this end, they should provide meaningful information, appropriate to the context, and consistent with the state of art:*
>
> - *to foster a general understanding of AI systems, including their capabilities and limitations,*
> - *to make stakeholders aware of their interactions with AI systems, including in the workplace,*
> - *where feasible and useful, to provide plain and easy-to-understand information on the sources of data/input, factors, processes and/or logic that led to the prediction, content, recommendation or decision, to enable those affected by an AI system to understand the output, and,*

[3] Office of Science and Technology.

[4] For example, Centers for Medicare and Medicaid Services (CMS) Health Plan Price Transparency Rules.

- *to provide information that enables those adversely affected by an AI system to challenge its output.*[5]

The Center for AI and Digital Policy's Universal Guidelines for AI say, "All individuals have the right to know the basis of an AI decision that concerns them. This includes access to the factors, the logic, and techniques that produced the outcome."[6] The Montreal Declaration for a Responsible Development of Artificial Intelligence couches transparency in its democratic participation principle. The document states that when governmental entities use AI, they must make the code publicly available.[7]

The National Institute of Standards and Technology (NIST) AI Risk Management Framework pairs accountability and transparency and notes that while other attributes are internal, accountability and transparency refer to both the "AI system and its external setting."[8] The NIST suggests that transparency concerns the availability of information. Transparency promotes understanding and is a prerequisite of accountability. The framework suggests that the type of information about a system should be tailored to the person interacting with the technology.

The EU requires trustworthy AI and has a risk-based framework. Transparency is a component of trustworthiness according to EU guidelines. Yet in medical settings, professionals using AI do not always meet transparency recommendations. A European study found that "Major transparency gaps included missing documentation on training data, ethical considerations, and limitations for deployment."[9]

While transparency may call for a sign off by which consumers or tech users note that information was made available to them, truly informed consent in medical decision making is something more. The language of informed consent is used beyond health care in settings where it may take on a consumeristic feel reflecting the principle of transparency.

2.2.1 Transparency in Practice

In institutions, transparency is an issue at the heart of governance. Employees and healthcare consumers deserve transparency. Transparency is an umbrella. Tech developers, those developing AI in-house, clinicians, and researchers may have different

[5] OECD AI Principles. Transparency and explainability. https://oecd.ai/en/dashboards/ai-principles/P7.

[6] Center for AI and Digital Policy. Universal Guidelines for AI. (2018) Website. https://www.caidp.org/universal-guidelines-for-ai/.

[7] Montreal Declaration for a Responsible Development of Artificial Intelligence. (2018) https://montrealdeclaration-responsibleai.com/.

[8] National Institute of Standards and Technology of the US Department of Commerce. https://nvlpubs.nist.gov/nistpubs/ai/NIST.AI.100-1.pdf.

[9] Fehr, J., Citro, B., Malpani, R., Lippert, C., & Madai, V. I. (2024). A trustworthy AI reality-check: the lack of transparency of artificial intelligence products in healthcare. *Frontiers in Digital Health*, 6, 1,267,290. https://doi.org/10.3389/fdgth.2024.1267290.

transparency obligations and different audiences to whom they are transparent. In hospitals and hospital systems, transparency about AI tools purchased, developed in-house, or in partnerships with biotech includes the following actions across the management's communications with clinicians and researchers:

- training medical practitioners to use AI properly,
- naming the benefits of the AI tool to those expected to use it,
- disclosing its limitations, including bias,
- describing the training data,
- allowing time for Q & A about each AI device,
- assigning a manager (IT, tech) to field technical questions, and
- explaining how data will be collected, used, and stored.

Before the AI arrives at the hospital, the AI developer or seller must be transparent with the purchaser. For example, the developer should disclose the following information:

- the basis for any accuracy claims (see Texas v. Pieces Technologies),[10]
- the risks associated with the technology, (medical and ethics related), including bias,
- the source of the training data,
- a description of how the AI was designed, including a basic outline of the algorithm and whether it uses machine learning,
- a list of any information omitted due to its being a protected trade secret or other intellectual property rights, and
- an explanation of how data will be collected, used, and stored. (Can the developer access the data? sell the data?)

Hospital information technology (IT) staff members or outside tech developers may be best positioned to explain how a tool was designed and to answer questions about it. Math, engineering, and computer science are increasingly relevant in health care.

Transparency about the use of AI is not a guarantee of the AI's accuracy, safety, or fairness. There are many open questions about how much information to disclose. For example, a consumer should be notified that data could be sold to a data broker, but should all possible outlets for the data be disclosed? Should consumers be surprised when advertisers seem to know their tastes and preferences very well?

It is important to raise public awareness about how AI is used in medical care. Transparency could promote a better understanding, decrease skepticism, increase trust, and ensure that a factual account of AI's uses, risks, and benefits is available. Transparency does not negate the need for informed consent. If anything, it can be viewed as a broader disclosure to the public, collectively to the users of an AI tool, an institution's healthcare consumer population, or the consumer.

[10] Texas v. Pieces Technologies, Inc. DC-24–13476, Dallas County, Texas 191st Judicial District. 2024. https://www.texasattorneygeneral.gov/sites/default/files/images/press/Petition%20for%20Approval%20of%20AVC%20Pieces%20File%20Stamped.pdf.

2.2.2 Transparency and Data

Consent to data collection and use has long been part of access to and use of technology. Consumers face and engage in consent processes often built into websites. Use of cookies is a common topic for which one provides a quick click, but not a signature. This announcement meets the requirements of transparency:

> This website uses cookies to enhance user experience and to analyze performance and traffic on our website. We may also share information about your use of our site with our social media, advertising and analytics partners.

This type of announcement or banner would then be followed by three buttons: accept cookies, reject all, and cookie settings. Consumers have a degree of control but cannot reject "required" cookies. The announcement adds transparency to something that used to be covert. The click does not meet the rigor of informed consent for several reasons. First, it is a quid pro quo that reflects a lopsided power distribution. In some jurisdictions, the operator of the website can prohibit the user from proceeding. If it is an online purchase or an informational website, the user is forced to agree to the terms or not use the website. This impacts voluntariness. Sometimes consumers need to access websites for government services, professional associations, research, jobs, and education. Additionally, many people are so used to seeing cookie banners that they do not even notice them. And the sign-off itself is not as rigorous as providing a signature would be.

2.2.3 Transparency and Trade Secrets

Intellectual property rights may conflict with transparency. They also may provide an excuse to be less transparent than one ought to be. Intellectual property rights provide incentives to create. Patents, copyrights, and trade secret protections play an important role so that others cannot piggyback on one institution's research and development or gain from the work of another. Yet limits on intellectual property are important as well. For example, food manufacturers must provide ingredient information. And, in health care, making generic drugs readily available provides a morally defensible argument against long periods of exclusivity. Transparency and intellectual property rights are often at odds—informed consent requirements may do a better job than the mere principle of transparency when it comes to the public's ability to access information about products including AI tools and pharmaceuticals.

2.3 How Transparency Differs from Explainability

While transparency and explainability are often described as a pair, their separation is preferable to ensure the ethical deployment of AI. Transparency runs human to human—sharing one's knowledge or uncertainty about a complex system. Transparency could involve disclosures about privacy, data collection, accuracy, bias, how the AI model works, the degree of explainability, and uncertainty. Transparency is a broadcast about AI or a disclosure directly to the consumer. While transparency and explainability are different concepts, there remains debate about how to be transparent when an AI tool is not explainable.[11]

Explainability refers to the AI model and the ability of a human to explain how it comes to its output. By definition, black box algorithms can't be explained by humans. However, the lack of explainability should not alone prohibit the use of black box AI models. AI that is not explainable may prove highly useful. The ability to sift through millions, and even billions, of images or data, and use them to evaluate an input, such as an image, means that AI could catch something the human eye would not. There is an ongoing tech ethics debate about the use of black box algorithms. Transparency about the nature of the system, its architecture, and a basic outline of its design may make a lack of explainability acceptable. However, there is no degree of transparency that would make unexplainable AI explainable.

There are also many tools designed to probe how AI comes to its recommendations (i.e., to crack the black box). These tools provide explanations designed for human understanding. For example, LIME (local interpretable model-agnostic explanations) creates a surrogate model[12] and SHAP (SHapley Additive exPlanations) uses game theory.[13] Local explanation tools seek to explain individual predictions while global explanation tools aim to explain the model. The growing market of tools to explain AI (the XAI market) is key to ethical deployment and XAI may improve the chances of regulatory approval as well. In a study about AI tools and stroke, researchers tested different XAI approaches finding valuable insights into how an otherwise unexplainable AI tool came to its output.[14]

Explainability as a principle leads to confusion. Explainability may be considered good, or even an ethical good. People and institutions have more trust in explainable AI. Explainability eliminates some uncertainty, improves a sense of understanding, and even empowers people to feel some control over or alliance with the AI tool. This does not imply that a lack of explainability is always bad or a type of ethical

---

[11] In the vast AI literature, some use transparency to refer to the AI itself, i.e., using transparency and explainability as synonyms. They are related but different concepts and here transparency describes human behavior and action rather than the AI tool.

[12] What is Local Interpretable Model-Agnostic Explanations (LIME)? C3AI. https://c3.ai/glossary/data-science/lime-local-interpretable-model-agnostic-explanations/.

[13] Welcome to the Shap documentation. https://shap.readthedocs.io/en/latest/.

[14] El-Geneedy, M., El-Din Moustafa, H., Khater, H. et al. (2025). A comprehensive explainable AI approach for enhancing transparency and interpretability in stroke prediction. *Scientific Reports* 15. https://doi.org/10.1038/s41598-025-11263-9.

failing. In healthcare, tools that are not explainable may be very helpful.[15] There may be a tradeoff between requiring explainability and innovation. In many cases, large datasets contribute to accuracy, reproducibility, and generalizability. Machine learning tools which make recommendations, classify, or categorize offer hope to those in need of solutions to medical problems.

Supporting an unexplainable AI tool with a medical theory may provide some comfort—a clinician's educated opinion about how an AI tool most likely came to a decision may offer comfort to the healthcare consumer. Such educated opinions should also take the AI model / architecture into consideration. For example, if AI picks up something in MRIs that is unnoticed by humans, a clinician, knowing some information about the dataset and algorithm, might hypothesize as to whether the AI has engaged in microscopic disease detection through visual analysis or has analyzed genetic data associated with high likelihood of disease progression and weighed the image and the genetics together.

Transparency goes beyond declaring that the tool is not explainable; clinicians should disclose what they do know about an unexplainable AI tool. Transparency around AI that is not explainable is important and should not be seen as a lost cause. Transparency about not only the basic architecture and the gist of the design, but also about the dataset, which problems the AI is meant to solve, and the percentage of accuracy it offers is crucial to its ethical deployment. For example, if AI makes a correct diagnosis 99.99% of the time and has gone through rigorous testing before being deployed, it is highly likely that the degree of accuracy is a better ethical floor than explainability, which would bar the use of highly accurate tools. If it were not tested enough or the dataset were too small or too homogenous, then it would not achieve the accuracy necessary to justify its broad application in health care, although it may be appropriate for some subgroups.

2.4 Transparency and Informed Consent Origins

Informed consent arose to counteract a paternalist tradition in medicine and resolve ethical violations in scientific research on human subjects. Its origin story is different from transparency's. Informed consent requires disclosures and therefore is consistent with transparency. The disclosures for informed consent are often more detailed and directed to the specific consumer of care rather than to the public or a group of consumers like information conveyed through a brochure. Transparency alone without specific rules and an acknowledgement by consumers or the public does not have the heft to prevent the harm that informed consent can prevent in the healthcare setting. Informed consent prevents unwanted touch or touch based on misunderstanding. The unwanted touch could be in the form of surgery or examination. For

[15] Miller, K. (2021) Should AI Models Be Explainable? That depends. *Stanford University Human-Centered Artificial Intelligence.* https://hai.stanford.edu/news/should-ai-models-be-explainable-depends.

example, gynecological training that involves touching a healthcare consumer who is under anesthesia without consent is condemnable. For a long time, many hospitals were not transparent about the practice, yet it was commonplace. Hospitals argued that training residents was an appropriate deviation from transparency and informed consent. Now, the United States Department of Health and Human Services (HHS) requires informed consent and will not allow Medicare or Medicaid billing if any such nonconsensual gynecological training exams take place.[16] Informed consent is more structured than transparency—it is not a principle; it is a requirement, a process, and should reflect a meeting of the minds. Informed consent as a practice enables healthcare consumers to make decisions based on correct information. It operationalizes the principle of autonomy. In some cases, it impacts the ability to attempt recourse when there is not both proper disclosure and a documented understanding of the risks.

Having distinguished transparency from informed consent, and transparency from explainability, the next chapters lay out the role of consent in the data use arena and then the evolution of informed consent to medical interventions.

[16] Marelli, A. (2024) New HHS guidance requires consent for pelvic, breast, and other sensitive examinations. *Stat News*. https://www.statnews.com/2024/04/01/informed-consent-pelvic-exam-new-hhs-guidelines/#.

Chapter 3
Data Landscape

3.1 Data Collection, Retention, Sale, and Use

AI requires massive amounts of data—it is used to train AI models; some data is collected when a person schedules an initial appointment; more data is collected throughout diagnosis, clinical trials, and treatment. EHR systems are complex and designed to share data among entities—within and among hospital systems, providers, insurers, the government, caregivers in some cases, and researchers. EHR systems include all sorts of consumer information from demographics to notes, symptoms, and prescriptions. The ethical issues are distinct: the use of data may in some cases bolster research and facilitate one's own treatment; yet in many cases, it will be lucrative for tech companies and hospital systems without contributing to either societal or personal wellbeing. Informed consent for data sharing is somewhat different from informed consent for medical procedures. One key difference is that the type of harm from privacy breaches, cybersecurity lapses, hacking, profiteering, and reidentification[1] is different from the physical harm associated with moving ahead with medical treatment without understanding the risks. For example, one harm may be stigma and discrimination associated with a disease. Another harm may be unjust enrichment in that hospitals make a lot of money selling data, but the subjects of the data are not compensated. Another important risk is a lack of cybersecurity leading to theft of social security numbers or passwords that enable cybercriminals. There is a push and pull between privacy, cybersecurity, and unjust enrichment on the one hand, and the desire to ease data sharing for training new AI models on the other.

---

[1] Reidentification has become increasingly possible due to technological advances. Sondeck, L., Laurent, M. (2025). Practical and ready-to-use methodology to assess the re-identification risk in anonymized datasets. *Scientific Reports* 15, 23223. https://doi.org/10.1038/s41598-025-04907-3 For an explainer on reidentification see Lubarsky, Re-Identification of "Anonymized" Data. (2017). Georgetown Technology Law Review. https://georgetownlawtechreview.org/re-identification-of-anonymized-data/GLTR-04-2017/.

A. Zimmerman, *Informed Consent for AI and Health Care*,
SpringerBriefs in Ethics, https://doi.org/10.1007/978-3-032-11347-4_3

Informed consent for data sharing and future uses is often built into consent forms in the research context. And within research, data sharing provisions are often broad to encourage data to be more freely shared, especially for future research. Some organizations providing research grants require broad data use provisions.[2] Now that the data routinely contributes to datasets that train AI, there are more opportunities to create AI products. However, healthcare consumers may not be aware of how their demographic, prescription drug, and treatment information may be shared.

3.2 Data Ownership

In AI ethics and law, there are many theories about data ownership and arguments over whether the data subject is de facto the data owner. I argue that the law should recognize individuals as owners of their personal data to an extent. Data ownership has proven difficult to define. Because data can be reused and concurrently used across systems, it is unlike physical property.[3] There are benefits to using the same data for many AI platforms—the plentifulness of data leads to more AI tools and may make each tool more accurate or better. For example, an abundance of individual data makes for excellent public health AI tools; an abundance of data also made generative AI tools possible and allows for higher quality creative output.

Under the European Union's (EU) General Data Protection Regulation (GDPR), the data subject seemed like an owner due to the many protections the law offers, including the right to know how data is collected and used and the right to correction and deletion. While the GDPR has some exceptions to the rights it contains, the general right is not waivable—individuals cannot trade away their data rights. Therefore, corporations processing and using data have obligations that individuals cannot relieve them of. The EU Data Act, becoming enforceable in September 2025, does not explicitly recognize ownership, but addresses uses and rights.[4] The Data Act encourages innovation and focuses on fair competitive terms to balance the power in the data landscape. The EU noted that many policymakers concur that ownership requires exclusion. The EU suggests that the reproduction and use of data doesn't impact the rights of the "original data holder." This is highly debatable because as data brokers buy datasets, they then own the dataset. Resale and reuse impact the data subject various ways. Data becomes less valuable to the data subject—a developer would not purchase data from the data subject if it were already available through third parties who sell it and profit from it. And, resale adds risk of reidentification,

[2] Representing Data Sharing in Informed Consent: Guidance for Researchers at the University of Illinois at Urbana-Champaign (2022). https://research.illinois.edu/sites/default/files/Representing%20Data%20Sharing%20in%20Informed%20Consent%20Guidance%20for%20Researchers_0.pdf.

[3] Graux, H. (2024) What is data ownership, and does it still matter under EU data law? European Commission. https://data.europa.eu/en/news-events/news/data-ownership-and-eu-law-examining-impact-data-act (data is inherently non-rivalrous, non-exclusive and inexhaustible).

[4] Graux, 2024.

privacy breaches, financial vulnerability, and extortion, etc. that the data subject continues to bear exclusively. The lack of a strong legal framework that imposes on big tech is a hindrance to developing and enforcing data ownership by data subjects. Furthermore, because data is often divulged freely, some suggest that rights to it are relinquished. The EU law is strong in that data subjects' rights persist after data is shared.

Data is also unlike intellectual property—personal information is raw data, not a creative endeavor. Intellectual property laws protect some databases as they are often curated content serving special purposes. Yet there is no intellectual property theory protecting most raw data. The ethical basis for protecting creative expression is not applicable.

Without acceptance of theories of data ownership, informed consent becomes an especially important tool to protect data subjects and allow them to have some control over how their data is used. Ownership would have advantages in that people would then be justified in seeking compensation for their contributions to research, AI training data, marketing tools, and commercial products. Absent ownership and the ability to track data as it is sold, resold, combined with other data, and separated from its original dataset, healthcare consumers are forced to donate their data. Many are pleased to do so for medical research. And the Health Insurance Portability and Accountability Act (HIPAA)[5] allows use of protected health information for research and nearly all uses of deidentified data without consent. Nonetheless, consumers often object to commercial use when given the option. Transparency and acknowledgement as either opt-in or opt-out consent already play prominent roles in the consumer data landscape.

3.3 Privacy and Data Use

The EU passed the GDPR in 2016 and it became effective in 2018.[6] The United States lacks a comprehensive data privacy law although some states have adopted laws to protect consumer data. California was the first state to adopt a consumer data protection law, the California Consumer Privacy Act (CCPA).[7] Now twenty states have their own laws.[8] Other states have narrow data protection laws, some pertaining to biometric or health data. Various federal laws address data and have

[5] 45 C.F.R. § 164.502 (1996).

[6] Regulation (EU) 2016/679 of the European Parliament and of the Council of 27 April 2016 on the protection of natural persons with regard to the processing of personal data and on the free movement of such data, and repealing Directive 95/46/EC (General Data Protection Regulation) (Text with EEA relevance) (OJ L 119 04.05.2016, p. 1, ELI: http://data.europa.eu/eli/reg/2016/679/oj).

[7] California Civil Code § 1798.100 et seq.

[8] Which States Have Consumer Data Privacy Laws? *Bloomberg Law.* (2025) https://pro.bloomberglaw.com/insights/privacy/state-privacy-legislation-tracker/.

enforcement mechanisms as well.[9] Many consumer data privacy laws are opt-out and do not require explicit consent for data collection and use. California's requires express consent for information about minors and for some uses of sensitive data. The laws tend to give consumers the right to know which data has been collected and for what purpose as well as some control over uses and rights to correction and deletion. CCPA has exceptions to deletion, for example entities do not need to honor consumer requests for deletion if the data is used to "Engage in public or peer-reviewed scientific, historical, or statistical research in the public interest that adheres to all other applicable ethics and privacy laws, when the business' deletion of the information is likely to render impossible or seriously impair the achievement of such research, if the consumer has provided informed consent."[10]

In research, the Common Rule, enacted in 1991, requires informed consent to the collection and use of identifiable data at any institution performing human subject research and receiving federal funding. The rule was codified and adopted across federal agencies engaging in research. Prior to a revision which took effect in 2019, it required institutional review board (IRB) approval or a fresh informed consent to use any identifiable data beyond the original use specified in the informed consent document. As the era of big data came about, deidentification became less foolproof and the possibility of reidentification has increased. The data landscape mixes swaths of data including consumer data, increasing the possibility of reidentification.[11] The 2019 Common Rule revision[12] watered down its consent requirement. Consent is now arguably both "less frequent" and "less informed."[13] The Revised Common Rule applies a reasonable person standard to disclosures about research, but it adopted broad consent. Broad consent allows future unspecified uses. While HHS justified the leap by the promise of big data in medical research, research subjects bear risks they did not under the previous rule. Having lost control over their data, research subjects may unwittingly contribute to research of which they morally disapprove. For example, in the health context, some oppose contributing their data to research about genetic anomalies or to train AI tools used for embryo selection.

Healthcare consumers provide massive amounts of data to hospital systems, pharmacies, and insurance companies. HIPAA facilitates data sharing, allowing insurers and other businesses to access medical records. State laws address some of the lapses in HIPAA and cover more entities. Consumer data laws have a larger reach covering the point of purchase and the data associated with many health- and healthcare-related products. Broad consent often applies to data collected through wearables

---

[9] HIPAA, COPPA, FERPA, FCRA, GLBA. (The FTC enforces laws and penalizes violators).

[10] California Civil Code § 1798.105 (d)(6).

[11] Froomkin, A.M. (2019) Big Data: Destroyer of Informed Consent 21 Yale J.L. & Tech. Special Issue 27, p. 48. https://yjolt.org/big-data-destroyer-informed-consent (quoting Nicholas Terry: "data triangulation may defeat any de-identification. In the second example, users increasingly generate wellness, fitness, and sickness data on mobile health platforms or by mobile health apps... Some data are created in a highly regulated space but then exported to a mobile device; other data are processed in the opposite direction").

[12] Revised Common Rule, § __.116(a)(5), 82 Fed. Reg. 7265/3 (codified at 42 C.F.R. § 46.116(a)(5)).

[13] Froomkin, 2019, p. 42.

like watches or through chatbots.[14] HIPAA is limited in that it governs providers, insurers, clearinghouses, plus business associates (third parties) though the Health Information Technology for Economic and Clinical Health (HITECH) Act.[15]

HIPAA does not restrict the use of deidentified data. The HIPAA privacy rule protects certain health information (protected health information) and governs which uses require consent. It is a relatively permissive law designed to promote "portability." The privacy rule has many exceptions. It permits disclosures of protected health information for many reasons beyond ease of insurance, clearing, and payment. For example, it permits sharing protected health information when the healthcare consumer is a victim of a crime, when releasing the data is required by law, for anti-money laundering purposes, when the release is court ordered or pursuant to a subpoena, to aid in criminal investigations, to report a crime, when the release furthers a national interest, public health interest, recall, or post-market issue, or contributes to communicable disease reporting, when decedents have an interest in it, with tissue/organ donation organizations, and in worker's compensation cases. HIPAA even permits sharing data for marketing, although that requires authorization.[16] HIPAA states, "A covered entity may use or disclose, without an individual's authorization, the psychotherapy notes, for its own training, and to defend itself in legal proceedings brought by the individual, for HHS to investigate or determine the covered entity's compliance with the Privacy Rules, to avert a serious and imminent threat to public health or safety, to a health oversight agency for lawful oversight of the originator of the psychotherapy notes, for the lawful activities of a coroner or medical examiner or as required by law." Covered entities are prohibited from requiring a healthcare consumer to sign an authorization to release data as a condition of receiving treatment or participating in a health plan.

Hospitals have so much data that they attract hackers.[17] There are risks of reidentification, identity theft, and exploitation. Some data collection is outside of informed consent. Opt-out provisions are a competing way of gathering consent. However, the line becomes blurred easily. For example, data related to inpatient food services delivered to the bedside has medical, personal, and consumer aspects.

Some suggest that medical research using data provided with the consent of the healthcare consumer is superior, or at least adequate, and that pouring time and

---

[14] National Academy of Medicine. 2025. An Artificial Intelligence Code of Conduct for Health and Medicine: Essential Guidance for Aligned Action. Washington, DC: The National Academies Press. https://doi.org/10.17226/29087, p. 29.

[15] 42 USC §§300jj et seq.; §§17901 et seq. (Prior to HITECH, HIPAA governed business associates based on contracts).

[16] HIPAA Privacy Rule (§45 CFR 164.501, 164.508, and 164.512[i]); https://www.hhs.gov/hipaa/for-professionals/privacy/laws-regulations/index.html.

[17] Rubio-Licht, N. (2025) Why Healthcare is a Hacker's Data 'Treasure Trove'. *The Daily Upside*. https://www.thedailyupside.com/cio/cybersecurity/why-healthcare-is-a-hackers-data-treasure-trove/.

money into research based on big data aggregating healthcare consumer and other consumer data is wasteful or not proven as miraculous as some researchers imply.[18]

Additionally, there is an ongoing tension between those in favor of repurposing as much data as possible for use in AI and those wanting consumers, publishers, and producers of intellectual property to control their data and creative works or share in the profits.[19] That debate invokes informed consent, concepts about data ownership, and rights to control data.

3.4 Unjust Enrichment

When it comes to data use, privacy isn't everything. Consumer data enriches corporations, and no profits or proceeds flow to the data subject. Hospital systems sell deidentified health data without consent.[20] As noted, that poses some risk of unjust enrichment but also creates some opportunity for reward in the form of medical research.

An old case somewhat sets the groundwork for the concept of unjust enrichment in the healthcare setting. In *Moore v. Regents of University of California*,[21] the plaintiff sued after the hospital system engaged in a lucrative pharmaceutical deal using his blood, bone marrow, and bodily fluid to create a cell line which was patented. The court validated the breach of fiduciary duty and breach of informed consent claims. Yet the court also held that the plaintiff did not have continued ownership of the blood, bone marrow, or the cells derived therefrom. The duty to disclose the use of bodily fluids is like a duty to disclose the use of data.

More recently, University of Chicago and Google created a research collaboration for the purpose of designing a new software product to decrease complications, eliminate unnecessary hospital stays, and improve health outcomes. The university provided Google with anonymized data (of specific healthcare consumers from a specific period) to be used as training data. In the contract between Google and University of Chicago, Google agreed not to reidentify any individuals whose data was part of the set.[22] In the case, *Dinerstein v. Google*, the plaintiff had been a patient

---

[18] Froomkin, 2019; Wetsman, N. (2021) Hospitals are selling treasure troves of medical data—what could go wrong? The Verge. https://www.theverge.com/2021/6/23/22547397/medical-records-health-data-hospitals-research (Eric Perakslis, chief science and digital officer at the Duke Clinical Research Institute said, "The usefulness of it is absolutely overblown").

[19] Longpre, S., Mahari, R., et al. (2024). Consent in Crisis: The Rapid Decline of the AI Data Commons. *Data Provenance Initiative* (MIT). https://www.dataprovenance.org/consent-in-crisis-paper.

[20] Wetsman, N. (2021) Hospitals are selling treasure troves of medical data—what could go wrong? The Verge. https://www.theverge.com/2021/6/23/22547397/medical-records-health-data-hospitals-research.

[21] 271 Cal.Rptr. 146 (1990).

[22] Dinerstein v. Google, LLC, No. 20-3134 (7th Cir. 2023). https://law.justia.com/cases/federal/appellate-courts/ca7/20-3134/20-3134-2023-07-11.html.

at the hospital, which collected "records of his vital readings, medical procedures, prescriptions, test results, and diagnoses. The records also contained demographic information." He sued claiming breach of contract, privacy violation, violation of the Illinois consumer fraud law, and unjust enrichment. In Illinois, there was no standalone tort of unjust enrichment. And without reidentification, there was no privacy violation or measurable financial or reputational harm to win the contract claim.[23] Therefore, it was impossible for the plaintiff to recover. However, in some jurisdictions, unjust enrichment cases are recognized as standalone torts. Bernard Chao suggests that because privacy harms are not usually economic, it is difficult to sue under current law.[24] Usually, financial harm is required. Rather than trying to fix those laws, plaintiffs should argue the common law tort of unjust enrichment.

Despite the failure of the plaintiff in Dinerstein to prevail, the concept of unjust enrichment is important. *It is inherently unfair that customers of any industry pay for goods and services with both their money and their data.* Unjust enrichment would require a showing that the defendant was enriched rather than a showing that the plaintiff suffered a financial loss.

As noted, the Revised Common Rule, which governs research on human subjects, now supports broad consent.[25] In the context of research, the NIH published recommendations for informed consent for future (secondary) research using data and biospecimens.[26] They presuppose like uses or perhaps similar disclosures for physical substances and data. Their document recognizes that corporations could profit from data and specimens. They recommend this sample language to organizations obtaining consent:

> The use of your data and biospecimens may lead to new tests, drugs, devices, or other products or services with commercial value. These products or services could be patented and licensed. There are no plans to provide any payment to you should this occur.[27]

Healthcare consumers often do not realize that their data may be sold and resold or be a key part of a lucrative tech or pharmaceutical development. It is difficult to assess how to convey this possibility in a meaningful disclosure. The blurb quoted above would not do the healthcare consumer any good. It would be insufficient to throw such important information into the fine print of a consent document. A clearer disclosure

---

23 Dinerstein v. Google (Additionally, the Court found a lack of standing on the Illinois Consumer Fraud and Deceptive Business Practices Act).

24 Chao, B. (2023). Unjust Enrichment: Standing Up for Privacy Rights and Privacy Losses as Wrongful Gains. *Iowa Law Review Online* Vol. 108:49 https://ilr.law.uiowa.edu/sites/ilr.law.uiowa.edu/files/2023-03/Chao_Final.pdf; Chao, B. Privacy Losses as Wrongful Gains, Iowa Law Review Online Vol. 106: 555 (2021). https://ilr.law.uiowa.edu/sites/ilr.law.uiowa.edu/files/2023-02/Chao.pdf.

25 Froomkin, A.M. (2019) Big Data: Destroyer of Informed Consent 21 Yale J.L. & Tech. Special Issue 27 https://yjolt.org/big-data-destroyer-informed-consent.

26 Office of Science Policy Office of Extramural Research, National Institutes of Health (2022) Informed Consent for Secondary Research with Data and Biospecimens: Points to Consider and Sample Language for Future Use and/or Sharing https://osp.od.nih.gov/wp-content/uploads/Informed-Consent-Resource-for-Secondary-Research-with-Data-and-Biospecimens.pdf.

27 Office of Science Policy Office of Extramural Research, National Institutes of Health, 2022.

on a separate document and with some verbal communication and expression of the possibility of lucrative pharmaceutical or AI products resulting from the data would be a better, higher ethical standard. Nonetheless, informed consent with any disclosure would serve not just to alert healthcare consumers to the possibility of profiteering that excludes them but also to alert collectively those most likely to take a stand against this common practice. A movement to either add a price tag to data or limit its use requires a knowledgeable consumer base or strict regulations.

In a stand against the unauthorized use of data, Italy found that OpenAI was unauthorized to scour the entire internet grabbing data to train ChatGPT.[28] Italy suggested OpenAI violated the GDPR by its mass collection and storage of data. The ethical issue of ChatGPT doing so can also be framed as unjust enrichment, a taking, and a violation of intellectual property rights.

[28] McCallum, S. (2023) ChatGPT banned in Italy over privacy concerns. *BBC News*. https://www.bbc.com/news/technology-65139406.

Chapter 4
Path to Informed Consent

4.1 Bioethics, Autonomy, and Informed Consent

In medicine, practitioners have a duty of care noted in the Hippocratic Oath,[1] the World Medical Association's Declaration of Geneva,[2] and American Medical Association (AMA) Code of Medical Ethics.[3] A duty to care is often associated with bioethical principles of beneficence and non-maleficence. Yet some bioethics literature contextualizes informed consent in a duty to care in that allowing for healthcare consumer autonomy is crucial to caring and thus to the ethical delivery of medical care. Traditionally, informed consent has been characterized as a right to know, a right to bodily integrity, and a right to information. It reflects autonomy, which itself is described as necessary to respect, dignity, virtue, freedom, independence, free will, self-defense, and self-preservation.

Bioethics literature has presented reasons for informed consent including promoting autonomy, promoting trust, combating paternalism, and protecting one's body.[4] Autonomy is the prevailing ethical basis for informed consent according to Beauchamp and Childress.[5] Kultgen frames consent as autonomy based but details

---

[1] Edelstein, L. (1943). *The Hippocratic Oath: Text, Translation, and Interpretation.* Johns Hopkins Press, 1943.

[2] Declaration of Geneva. World Medical Association. (2024). https://www.wma.net/policies-post/wma-declaration-of-geneva/ ("the health and well-being of my patient will be my first consideration").

[3] (duty to treat).

[4] Cohen, I.G. (2019). Informed Consent and Medical Artificial Intelligence: What to Tell the Patient? *Georgetown Law Journal*, Vol. 108. https://www.law.georgetown.edu/georgetown-law-journal/wp-content/uploads/sites/26/2020/06/Cohen_Informed-Consent-and-Medical-Artificial-Intelligence-What-to-Tell-the-Patient.pdf (Cohen refers to patient protection, autonomy, trust, personal integrity, self-ownership, and non-domination as six normative justifications for informed consent).

[5] Beauchamp, T. and Childress, J. (2019). *Principles of Biomedical Ethics. Eighth Ed.* Oxford: Oxford University Press.

A. Zimmerman, *Informed Consent for AI and Health Care*,
SpringerBriefs in Ethics, https://doi.org/10.1007/978-3-032-11347-4_4

its limitations. Those wishing for another person's consent must "persuade him rationally—that is, to secure his rational and informed consent."[6] Consent is framed in the context of *doing for* someone else. It is the strongest when the person consenting is fully informed.[7] To Kultgen, consent would not ethically justify a bad act. The person acting *on behalf of* the person who consents must exercise discretion. Framing the clinician as both agent and protector is noteworthy in that rather than seeing informed consent as a sacrifice of autonomy, it can be framed as a way for the healthcare consumer to attain a good that requires resources and help, i.e., a good that the consumer cannot access independently.

Consent goes further than handing documents off for signature—the task of rationally persuading often requires verbal explanations and the hard work of building a trusting relationship. Onora O'Neill has argued for a trust-based rationale, suggesting that informed consent requirements are important for building and maintaining trust in practitioners and institutions.[8] Additionally, some justify informed consent as fortifying against assault, deceit, coercion, and exploitation.[9]

Despite both clinician and healthcare consumer sharing medical goals, conflicts are not uncommon in medicine. In research, arguably the goals are not as shared: an individual could become a research subject for money rather than in hopes of a specific medical outcome. Areas of disagreement about informed consent include which information must be disclosed (adequacy), management and meaningfulness of consent when there are not choices of treatment, and voluntariness.

4.2 Informed Consent as Ethics and Law

The ethical requirement of informed consent is arguably broader than the legal requirement—ethics may call for instituting informed consent in circumstances beyond those that the law requires. Like a contract, informed consent requires a meeting of the minds.[10] Consent entered involuntarily or under duress is not valid. Informed consent is a legal and ethical obligation. Case law is helpful in setting some minimum requirements, while an inquiry about whether disclosure is ethically required goes beyond cases. Ethically speaking, one factor it depends on is whether

---

[6] Kultgen, J. (1995). Autonomy and Intervention: Parentalism in the Caring Life. Oxford University Press.

[7] Kultgen, 1995.

[8] O'Neill, O. 2002. *Autonomy and Trust in Bioethics*. Cambridge: Cambridge University Press.2002, 145.

[9] Manson, N., and O'Neill, O. 2007. *Rethinking Informed Consent in Bioethics*. Cambridge: Cambridge University Press. (dismissing informed consent as abuse prevention).

[10] Courts have found that consent documents are a contract. See Resnik D. B. (2009). Do informed consent documents matter? *Contemporary Clinical Trials, 30*(2), 114–115. https://doi.org/10.1016/j.cct.2008.10.004; there are mixed views on this; in many jurisdictions, including New York, informed consent documents are not a contract.

the consumer of medical care would use the information to make a decision. If so, the disclosure should be ethically compelled. This is a question of materiality.

Battery was the classic cause of action for an alleged breach of informed consent—nonconsensual touch. A breach of informed consent can constitute medical malpractice, although in some jurisdictions it is a standalone cause of action. Informed consent case law demonstrates the risk of failing to inform the consumer of medical care. This is highly relevant in the case of AI. Let's say a healthcare consumer agreed to a surgery based on an AI diagnosis or recommendation without knowing that the clinician did not challenge the AI, consider and present all reasonable options, or use any resources other than AI. If the person later learns that a less intrusive treatment that the algorithm did not recommend might have worked, there could be a valid claim. As clinicians increase their use of clinical decision support tools, even using them as a replacement for their own expertise, the scenario is likely to become commonplace. In other settings, deference to automation has led to humans questioning their own expertise or assessment.[11]

Informed consent is a process, not merely signed forms. Ideally, in the context of disease care, informed consent begins with conversations about diagnosis and treatment options. To consent to one intervention often requires knowledge about other options. The consent forms for complex care follow important conversations after which an intervention is decided upon.

Elements of informed consent include disclosure (by the practitioner), understanding (by the consumer of care) of information, *and* a voluntary sign off. It is somewhat a permission slip. Informed consent is a guardrail to ensure autonomy and an expression or documentation of an autonomous decision. The degree of specificity is often debated because some details are ancillary to decision making, while others are crucial. Often ancillary actions, like surgical prep, are not specifically described in consent forms, yet their consent is presumed. These three requirements (disclosure by clinicians, understanding, and voluntary consent by the healthcare consumer) must be met for medical procedures absent special circumstances like an unconscious person experiencing a medical emergency. The National Institutes of Health (NIH) says informed consent must be obtained before surgeries, complex medical tests and procedures, such as an endoscopy, needle biopsy, radiation or chemotherapy, genetic testing, vaccines and blood tests.[12] There are codified requirements for informed consent in medical research. HHS lays out those requirements,[13] which were designed to protect research subjects from abusive human research practices.

[11] Eubanks, V. (2018). *Automating Inequality*. Picador. (Provides example of algorithms' outsized influence on child services screeners and caseworkers, who then question their own judgment).

[12] MedlinePlus (US Government) https://medlineplus.gov/ency/patientinstructions/000445.htm; United States Department of Health and Human Services website. Informed consent. www.hhs.gov/ohrp/regulations-and-policy/guidance/informed-consent/index.html. Updated August 31, 2023.

[13] 45 CFR §46.116.

4.3 Legal History

The ability of healthcare consumers to weigh the type and degree of risk of a medical intervention is crucial to making the decision to go ahead with or to deny the treatment. A personal judgment about safety requires some information about the safety of the treatment. Personal risk assessment generally includes one's own priorities, including personal opinions, socioeconomics, and obligations unrelated to medicine, etc. Whether the use of AI is relevant depends on the AI itself, how it is used, and the type of medical decision, among many other factors. Its relevance to consumers depends on whether and how it would or could impact medical decision making and treatment. Courts often look to materiality in assessing informed consent breaches. The ethical disclosure of AI may depend on other factors.

A 1767 English court ruled that physicians must obtain patient consent prior to an "operation."[14] In the United States, in Pratt v. Davis,[15] a 1905 case, an Illinois court noted the following:

> Under a free government at least, the free citizen's first and greatest right, which underlies all others—the right to the inviolability of his person, in other words, his right to himself—is the subject of universal acquiescence, and this right necessarily forbids a physician or surgeon, however skillful or eminent, who has been asked to examine, diagnose, advise and prescribe (which are at least necessary first steps in treatment and care), to violate without permission the bodily integrity of his patient by a major or capital operation, placing him under anaesthetics for that purpose, and operating on him without his consent or knowledge.[16]

Pratt v. Davis (1905)
Facts: Person with epilepsy had cervical/uterine surgery. Doctor gave less than full disclosure to get consent—he said the person needed a small surgery on the "neck of the womb." The doctor then performed major surgery.
Holding: Without consent, "… a surgeon has no right to violate the person of the patient by a serious major operation or one removing an important part of the body."

In Mohr v. Williams,[17] a doctor determined a person's right ear was diseased and required surgery. However, once the patient was anesthetized, the doctor found the left ear to be in worse condition and operated on that. The patient had not been informed of any problem with the left ear. Some hearing loss resulted. The person sued for assault and battery and initially won a sizable verdict, which was later set aside as excessive, considering the doctor's goodwill. However, consenting to surgery on one ear does not equate to an implied consent to surgery on the other ear.

---

[14] Emanuel, Z., and Steven Joffe, Informed Consent, in Holland-Frei Cancer Medicine, 6th Edition. https://www.ncbi.nlm.nih.gov/books/NBK12636/ (court relies on tradition to hold physician accountable).

[15] Pratt v. Davis. 224 Ill. 300 (1905).

[16] Pratt v. Davis, 1905.

[17] Mohr v. Williams, 95 Minn. 261 (1905).

Mohr v. Williams (1905)
Facts: After the person agreed to surgery on the right ear, the doctor operated on the left ear instead because the left ear was in a worse condition. This was not an emergency.
Holding: Absent special circumstances, physicians performing surgery without consent are acting unlawfully.

The era of informed consent in the United States became more widely accepted after the 1914 Schloendorff[18] case in which a patient specifically refused surgical removal of a tumor when she approved an exploratory surgery. She awoke to find her fibroid tumor removed. Later, gangrene developed in her left arm, and she had to resort to having her fingers amputated. Schloendorff sued for battery, assault, and trespass. The famous quote, "Every human being of adult years and sound mind has a right to determine what shall be done with his own body," frames informed consent as a rights-based notion, and exercising informed consent or refusal as an act of autonomy.[19] Schloendorff was not as much about the appropriate disclosure—not about the *informed* part of informed consent—but the sign-off. The patient specifically said not to remove anything, just to look.

Schloendorff v. Society of New York Hospitals (1914)
Facts: Patient instructed doctor only to explore, not remove. Doctor removed fibroid tumor. Post-surgical gangrene led to finger amputation.
Holding: Operating without consent violates the right to bodily autonomy.

The early cases firmed up the concept of autonomy. Then, case law began to fine tune which disclosures satisfy the demands of informed consent.

The more serious or common the side effects, the stronger the need for disclosure. If viewed as a sliding scale, rare but very serious side effects require disclosure[20]; rare yet insignificant ones likely do not. Informed consent arose out of a combination of healthcare consumers' rights and safety. There are two standards commonly applied to informed consent cases turning on disclosure. One standard is referred to as the reasonable person standard (also called the reasonably prudent person standard and the materiality standard). The other is the physician standard (also called the customary standard and the community standard).

---

18 Schloendorff v. Society of New York Hospitals, 211 N.Y. 125 (1914)(Schloendorff is also commonly noted for its prohibition against suing charitable hospitals for the action of the doctor performing surgery there).

19 Schloendorff, 1914 (held that a charitable hospital cannot be held responsible for the acts of a physician).

20 Salgo v. Leland Stanford Jr. Board of Trustees, 154 Cal. App. 2d 560 (1957) (no warning of risk of paralysis; physician liable for withholding information); Canterbury v. Spence 464 F.2d 772(D.C. Cir.), cert. denied, 409 U.S. 1064 (1972) (physician must disclose all risks which a reasonable person would consider significant in making the decision to proceed; failure to disclose chance of paralysis violated informed consent—the consent was not informed; case is noted for introducing the reasonable patient standard and not allowing the medical community's standard to apply and bar a medical malpractice lawsuit).

Salgo v. Leland Stanford Jr. University Board of Trustees (1957) established some disclosure requirements. An aortogram left the plaintiff paralyzed. He sued arguing he was not aware of the risk of paralysis. The jury was instructed that a physician must disclose to the healthcare consumer "all the facts which mutually affect his rights and interests and of the surgical risk, hazard and danger, if any…"[21] The court noted that the physician cannot "minimize any known dangers", yet the court cautioned about over-disclosure, suggesting that physicians should use their discretion to avoid alarming patients.[22]

Salgo v. Leland Stanford (1957)
Facts: An aortogram caused paralysis.
Holding: Failure to disclose risk is a proper basis for a cause of action.

In *Canterbury v. Spence*,[23] the DC Circuit Court of Appeals weighed in on informed consent. A patient had back surgery and a subsequent fall, resulting in bowel paralysis, incontinence, and a need for crutches to walk. The court stated that the relevant standard is *not what is customary, but what is reasonable*. One question based on *Canterbury v. Spence* is informative: Would a reasonable person consider the information significant when deciding whether to engage in or refuse medical treatment? *Canterbury v. Spence* clarified the disclosure requirement and identified disclosure as crucial to the meaningfulness of self-determination, i.e., autonomy. The court applied the common objective standard used in torts rather than continuing a special rule for physicians. "[W]e agree that '[a] risk is thus material when a reasonable person, in what the physician knows or should know to be the patient's position, would be likely to attach significance to the risk or cluster of risks in deciding whether or not to forgo the proposed therapy.'".

Canterbury v. Spence (1972)
Facts: Laminectomy followed by a fall led to paralysis and chronic disability.
Holding: Physician has a duty to disclose based on what a reasonably prudent person would find significant when deciding to provide consent to a treatment. The standard of physician tradition or custom is not to be applied. (Yet the reasonable person standard as applied depended on the physician's view of the reasonable patient.)
Interpretation: Physicians must reveal risks of severe consequences even if remote or rare.

Some informed consent lawsuits address proper management of the healthcare consumers' expectations. For example, misunderstandings (overestimates) of the chance that chemotherapy would be curative may lead people to agree to it. In *Arato v. Avedon*, a 1993 case, the patient had pancreatic cancer, which had a five percent five-year survival rate. The physician did not disclose the low odds to him. The patient's wife sued, arguing had the patient understood the low probability of survival, he

[21] Salgo, 1957 (discussion of duty to disclose).
[22] Salgo, 1957 (recommending jury be told physicians can use discretion about disclosure).
[23] Canterbury v. Spence 464 F.2d 772 (D.C. Cir. 1972).

would have wished to die peacefully at home.[24] The court noted that the duty to disclose "is not governed by the standard practice of the physician's community."

Arato v. Avedon (1993)
Facts: Pancreatic cancer. Doctor did not reveal the low probability of cure.
Holding: Doctor must disclose material information so patient can make an informed choice.

Disclosure of a physician's experience (amount of time practicing or using a certain procedure or device) is an area in which informed consent is evolving.[25] In *Johnson v. Kokemoor*,[26] a 1996 Wisconsin case, neurosurgeon Dr. Kokemoor had performed the clipping of two basilar bifurcation aneurysms but implied he had performed many. He had done more than 30 anterior circulation aneurysms. Prior to her diagnosis, Johnson's only symptom was that she was experiencing headaches. Surgery rendered her partially quadriplegic and unable to walk or control her bowels and bladder. The issue in the case was whether the law, which required disclosure of risks and benefits, required disclosure of inexperience. The court ruled that a reasonable person would consider a physician's surgical experience in deciding whether to go ahead with surgery.[27] *Johnson v. Kokemoor* is an informed consent case—the judge noted that the issue was whether a reasonable person in the plaintiff's position would not have consented to the surgery if she had the pertinent information. (Because the claim was not malpractice, the question of whether the doctor was negligent or should have referred the case to a more experienced surgeon was not appropriate for the jury.)

Johnson v. Kokemoor (1996)
Facts: Clipping basilar bifurcation aneurysm caused significant irreversible harm.
Holding: The surgeon's level of inexperience was material.

In *Montgomery v. Lanarkshire Health Board*,[28] the UK's Supreme Court shifted the disclosure requirement there, creating a reasonable patient standard for evaluating whether the information would be material. The Court had previously applied a standard based on whether a reasonably prudent physician would have disclosed the information. This movement away from paternalism in medicine in the UK supports patient autonomy. In the case, a small woman was not told about the risks of vaginal delivery with diabetes. Her baby was born with shoulder dystocia and disabilities.

---

[24] Arato v. Avedon, 23 Cal. Rptr. 2d 131 (1993).

[25] Banja, J. (2015) Disclosure of Experience as a Risk Factor in Informed Consent for Neurosurgery: The Case of Johnson v. Kokemoor. Virtual Mentor. Vol. 17(1):69–73. https://doi.org/10.1001/virtualmentor.2015.17.1.oped1-1501.

[26] Johnson v. Kokemoor, 199 Wis2d 615, 545 NW2d 495 (1996).

[27] Johnson v. Kokemoor, 199 Wis2d 615, 545 NW2d 495 (1996). https://law.justia.com/cases/wisconsin/supreme-court/1996/16883.html (reversing court of appeals and remanding; requiring disclosure of material facts for informed decision making); see also Johnson v. Kokemoor, 188 Wis. 2d 202 (1994), reversed, ("the duty to inform does not include a duty to advise of and refer to better surgeons and medical centers").

[28] Montgomery v. Lanarkshire Health Board. UKSC 11 (2015).

The court determined that, "The test of materiality is whether in the circumstances of the particular case the court is satisfied that a reasonable person in the patient's position would be likely to attach significance to the risk."[29]

Montgomery v. Lanarkshire Health Board (UK, 2015)
Facts: Vaginal delivery with diabetes caused complications the risk of which the woman was unaware.
Holding: Doctor must inform the patient of all material risks—courts will apply a reasonable patient standard.

In *Allen v. Harrison*, a 2016 Oklahoma case,[30] a person swallowed a nail and went to the hospital with the nail in her stomach. Rather than remove it, the doctor told her to "eat fiber" and allow it to pass through. She developed an infection that required surgery. She brought a medical malpractice suit based partly on a breach of informed consent. One issue was whether informed consent is needed for noninvasive options. The secondary issue was whether the doctor should have disclosed alternatives, including endoscopic or other surgery to remove the nail.

Allen v. Harrison (2016)
Facts: Doctor did not disclose the option of removing nail from stomach.
Holding: Informed consent is required for nonsurgical course of treatment; and physicians must discuss reasonable alternatives even if not recommending them.

In *Davies v. MultiCare Health Systems*,[31] the plaintiff was in a car accident. She went to the emergency room where she was diagnosed with a spinal fracture requiring a collar. She was then discharged. The next day, she had a stroke. The accident had caused a vertebral artery dissection which caused the stroke. Davies claimed she was never advised about the risk of vertebral artery dissection. Treating it would likely have prevented her stroke. The doctor did not note any symptoms of vertebral artery dissection. Whether the doctor was negligent in failing to order diagnostic tests was separate from the informed consent claim. The court held that a doctor cannot have a duty to disclose that which he did not diagnose, even in the case of misdiagnosis. "But she cannot bring an informed consent claim where, as here, the physician ruled out the undiagnosed condition entirely. The physician is not required to obtain informed consent about taking further action to detect or mitigate a condition that the physician has ruled out." In its holding, the court clearly split informed consent cases from medical malpractice cases. The plaintiff also sued for negligence. The medical community felt this was a favorable decision.[32]

Davies v. MultiCare Health System (2022)

---

[29] Montgomery v. Lanarkshire Health Board. UKSC 11 (2015).

[30] Allen v. Harrison, 374 P. 3d 812 (Okla. 2016).

[31] Davies v. MultiCare Health System. Davies v. MultiCare Health Sys., 510 P.3d 346 (Wash. 2022).

[32] Schlicher, N. (2022). Davies v. Multicare: Supreme Court Gets It Right. *Washington State Medical Association.* https://wsma.org/Shared_Content/News/Membership_Memo/2022/June_24/davies_v_multicare_supreme_court_gets_it_right.

Facts: Plaintiff had a spinal injury and was sent home; had a stroke due to undiagnosed vertebral artery dissection.
Holding: Informed consent and medical malpractice are distinct causes of action; no duty to disclose risks associated with an undiagnosed, unsuspected injury; medical malpractice claim may be appropriate depending on negligence.

A reasonable healthcare consumer standard applied uniformly would clarify informed consent. At least twenty three states and Washington, DC have a healthcare consumer-based materiality standard.[33] Many jurisdictions have codified informed consent disclosures.[34] For example, in Washington state, informed consent claims require a showing that a "a reasonably prudent patient under similar circumstances would not have consented to the treatment if informed of such material fact or facts[.]"[35] New York's law states that, to recover, a plaintiff must show that a "reasonably prudent person in the patient's position would not have undergone the treatment or diagnosis if he had been fully informed and that the lack of informed consent is a proximate cause of the injury or condition for which recovery is sought."[36]

When evaluating whether to disclose information, clinicians should note that the disclosures for informed consent do not need to be outcome determinative individually in healthcare consumer decision making. For example, in the context of chemotherapy, one person deciding whether to consent or refuse a drug may not care about hair loss while another would. The one who does not care as much about hair loss may be extremely opposed to hearing loss, for example, while another healthcare consumer may be fearful of daily nausea. A reasonable healthcare consumer standard is objective, not individual. Courts would likely find that a reasonably prudent person would care about hearing loss and hair loss and that to fail to disclose either one would be a breach of informed consent.

Nonetheless, many jurisdictions continue to apply a physician-based (community) standard rather than a reasonable person standard. They favor the customs of physicians or what physicians deem reasonable. The physician-based standard generally asks what other physicians in similar circumstances and in the same state would disclose.[37] The standard derives from custom: if the industry custom is lax and less is disclosed, the lower standard would absolve the physician. Therefore, the physician-based standard is usually characterized as less healthcare-consumer friendly, and, importantly, less plaintiff friendly. Many states including Alaska,

---

[33] Weinmeyer, R. (2014). Lack of Standardized Informed Consent Practices and Medical Malpractice. *AMA Journal of Ethics*. https://journalofethics.ama-assn.org/article/lack-standardized-informed-consent-practices-and-medical-malpractice/2014-02; Spatz, E. S., Krumholz, H. M., & Moulton, B. W. (2016). The New Era of Informed Consent: Getting to a Reasonable-Patient Standard Through Shared Decision Making. *JAMA*, 315(19), 2063–2064. https://doi.org/10.1001/jama.2016.3070.

[34] Paterick, T., Carson, G., et al. (2008). Medical Informed Consent: General Considerations for Physicians. Mayo Clinic Proceedings. https://www.mayoclinicproceedings.org/article/S0025-6196(1160864-1/fulltext.

[35] Wash. Rev. Code § 7.70.050.

[36] New York Pub. H. §2805-d.

[37] Weinmeyer, 2014.

Arkansas, Delaware, Florida, and Kentucky have codified physician-based standards for disclosure. Yet, ethically speaking, as outlined in Chapter 5, clinicians should disclose what a reasonable healthcare consumer would want to know.

Wisconsin codified its reasonable-physician standard: "The reasonable physician standard requires disclosure only of information that a reasonable physician in the same or a similar medical specialty would know and disclose under the circumstances." The same statute goes on to note that physicians need not disclose "Detailed technical information that in all probability a patient would not understand."[38] This is especially relevant to the use of AI in health care. Such statutory language may provide an excuse for those failing to disclose material technical information. A low bar for the law need not indicate that the ethical bar should be low as well.

In medicine, informed consent is often intertwined with capacity assessment and legal competence, both beyond the scope here. Clinicians try to judge whether consumers of care understand their health circumstances, treatment options, and prognosis. Some degree of understanding is crucial to providing consent. AI poses some issues as many people do not understand AI very well—yet clinicians should aim to describe AI in plain language with a focus on both its purpose and its architecture.

[38] Wisconsin Statutes 448.30 Informed consent.

Chapter 5
Reasons to Disclose the Use of AI

5.1 Ethics of Informed Consent for Healthcare AI

As noted, the accepted ethical rationale for informed consent is autonomy.[1] To ensure that consumers of care can make autonomous decisions, clinicians must provide the information necessary upon which healthcare consumers base their informed consent. The physician standard is less healthcare-consumer-centered than the reasonable person standard, which would tend to favor healthcare consumers. Applying ethics, rather than law, yet drawing on law, there are many reasons to supply the information consumers may use in making healthcare choices. Once disclosed, a meeting of the minds should follow. To meet the requirements of ethics in health care, clinicians should err on the side of disclosure and avoid using some jurisdictions' lax informed consent laws as an excuse to fail to disclose meaningful information. By type or degree of risk, the middle ground is where the ethical issues will arise.

5.2 Why Disclose the Use of AI and Obtain Informed Consent?

Informed consent is the most consumer-friendly, autonomy-respecting framework for AI in health care. An informed consent framework for AI is practical for various reasons:

- informed consent is already built into the healthcare setting,

[1] Beauchamp, T. and Faden, R. (2014) Informed Consent, in *Bioethics, 4th Ed.* Online Cengage Learning. https://bioethics.jhu.edu/wp-content/uploads/2022/05/Beauchamp-TL-Faden-RR-Meaning-and-Elements-of-Informed-Consent.pdf And see Daniel Callahan, D. (1984) "Autonomy: A Moral Good, Not a Moral Obsession," *The Hastings Center Report* 14, no. 5 (1984): 40–42, https://doi.org/10.2307/3561098.

A. Zimmerman, *Informed Consent for AI and Health Care*,
SpringerBriefs in Ethics, https://doi.org/10.1007/978-3-032-11347-4_5

- it empowers the healthcare consumer and meets the requirements of autonomy,
- it is anti-paternalism, and
- AI poses many of the same ethics-related risks as already seen and evaluated in bioethics and medical practice.

Therefore, exploring when and whether AI is material to decision making is imperative. While AI is unlike other tools in health care, informed consent has been used to document risks such as safety, privacy, limited reliability, low probability of cure, medical error, and misdiagnosis. From traditional surgeries to complex treatments, informed consent has been the technical tool through which healthcare consumers exercise autonomy.

To summarize what is detailed below, the widespread use of informed consent for AI in health care is called for because healthcare consumers do want to know when and how AI is used, it is reasonably likely to impact their decision-making process, i.e., it is often material, and they do not always know to ask about it. The use of AI may impact the healthcare consumer's perception and the weight given to a physician's recommendation.[2] Clinicians may overly rely on AI, making it more likely to impact their medical recommendations. A failure to obtain informed consent could be deceptive and legally risky. Healthcare consumers' rights to refuse care and to have control over their bodies are fundamental human rights.[3] Autonomy may foster trust—failing to obtain informed consent when using AI in health care could reduce trust in healthcare institutions, clinicians, and researchers.

5.2.1 Reasonable People Do Want to Know About the Use of AI

A Pew Research survey revealed that 60% of healthcare consumers would be "uncomfortable with the doctor relying on AI" in their health care.[4] Healthcare consumers do care whether and how clinicians use AI. The public is also nearly evenly split—some believe AI will lead to better "health outcomes" and others do not. More people believe it would reduce mistakes than contribute to them. And many believe AI will hurt the doctor-patient relationship. Another pronounced concern is that doctors will adopt new technologies too quickly rather than using appropriate caution. The responses indicate a concern that hospitals will adopt new technologies "before fully understanding the risks to patients."[5]

[2] Findley, et al., 2020.

[3] UN Declaration on Bioethics and Human Rights. UNESCO, *Ethics of Science and Technology.://* www.unesco.org/en/ethics-science-technology/bioethics-and-human-rights.

[4] Tyson, A., Pasquini, G., and Spencer, A., and Funk, C. (2023). 60% of Americans Would Be Uncomfortable With Provider Relying on AI in Their Own Health Care. Pew Research Center. https://www.pewresearch.org/science/2023/02/22/60-of-americans-would-be-uncomfortable-with-provider-relying-on-ai-in-their-own-health-care/.

[5] Tyson, et al., 2023.

Another poll also indicated that healthcare consumers want to know about the use of AI. Patients responded that doctors should always disclose the use of AI in answering patient texts (84%), analyzing lab results (74%), triage (71%), patient education (69%), creating EHR notes (68%), writing research papers (64%), and sharing information (61%). Slightly fewer than 50% responded that physicians should always disclose the use of AI in accessing medical research and submitting insurance claims.[6]

Additional studies suggest that AI makes information seem more important. Healthcare consumers perceive diagnoses based on AI as important to their decision making.[7] Because they value sources of information differently, knowing where the information came from is important to them. AI impacts decisions and is important and material.

Not only do consumers want to know whether and how AI is used in their care, AI has some risks while it can minimize other risks. Many AI tools identify correlation. Consumers may wish to know about the absence of a causal relationship because that may be material to their decision making. AI can be prone to some known errors or programmed to err on the side of false positives rather than false negatives, or vice versa. Consumers have an interest in knowing the basics about the tools used to diagnose and treat.

Generative AI is important to healthcare consumers as personal communications impact the doctor-patient relationship. Generative AI is becoming more sophisticated and can convey empathetic responses.[8] A stronger "higher quality" response may come at the expense of sincerity.[9] Consumers of care following up on AI-generated communications in person at appointments risk finding clinicians have not read over the AI-generated conversation. In the same vein as the Kokemoor case, some consumers of care would choose a different clinician. That is, communication AI may be material to the decision about where to continue care just as clinician experience may be. The decision to continue medical care based on the thoroughness of responses that the consumer thought were personal would be based on deception or misunderstanding. AI in communications can impact healthcare decision making more directly as consumers would be basing healthcare decisions on some combination of the clinician and AI communication.

AI may be harmful, depending on the scenario. There are inherent risks, including biased data, a rogue unexplainable algorithm, an error of accuracy, generalizability,

[6] Pearl, R. Poll: AI disclosures between doctors, patients. Robert Pearl, MD. https://robertpearlmd.com/poll-ai-disclosures/ (poll of 511 people).

[7] Park H. J. (2024). Patient perspectives on informed consent for medical AI: A web-based experiment. *Digital health, 10*, 20552076241247938. https://doi.org/10.1177/20552076241247938.

[8] Tai-Seale, M., Baxter, S., Vaida, F., et al. (2024). AI-Generated Draft Replies Integrated into Health Records and Physicians' Electronic Communication. *JAMA Netw Open*. https://doi.org/10.1001/jamanetworkopen.2024.6565AI; Pierce, A., (2024) Study Reveals AI Enhances Physician–Patient Communication. *UC San Diego Health*, press release. https://health.ucsd.edu/news/press-releases/2024-04-15-study-reveals-ai-enhances-physician-patient-communication/.

[9] Tai-Seale, et al. 2024. (The study of AI-generated replies given through EPIC did disclose the use of AI to the consumers of care).

or reliability, distraction and deviation from the gut instinct and the expertise of the clinician, and medical error. As noted, additional risks include privacy violations and cybersecurity breaches which could lead to reputational as well as financial harm. Because AI poses risks and physicians are well positioned to understand the risks posed by tools they use, its use should be disclosed. Applying a reasonable person standard, it is highly likely that many healthcare consumers would find the use of AI material when making a decision to proceed with or refuse treatment.

5.2.2 *Healthcare Consumers May not Know to Ask*

Some argue that healthcare consumers who ask about whether AI is used deserve an honest answer.[10] And, of course, they do. However, providing information about AI only to those who request it is a bare minimum and would leave many healthcare consumers in the dark. Healthcare consumers vary in their knowledge, suspicion, savviness, and preferences. Many may not naturally suspect AI's use in their medical care. Ignorance is not bliss—healthcare consumers who do not know about the use of AI have no reason to ask about it.

In the *Practice Fusion* case, Purdue Pharma and other companies had paid Practice Fusion to have the EHR system issue popup alerts recommending long-release OxyContin and other drugs.[11] While the scheme violated the Anti-Kickback Statute and the False Claims Act, consumers did not know their doctor's suggestion was based primarily on an EHR popup. It is unlikely any healthcare consumer learned that the popup was the reason that a clinician prescribed the addictive drug. Christina Nolan, US Attorney for the District of Vermont, noted the practice allowed "the drug company to have its thumb on the scale at precisely the moment a doctor was making incredibly intimate, personal, and important decisions about a patient's medical care, including the need for pain medication and prescription amounts."[12]

Generative AI may engage in "hallucination" in written materials, images, code, etc. If consumers do not know AI is used, they may overlook or disregard an anomaly in their medical record, report, or communication which they would have questioned if they had known about the use of AI. This of course depends on which they trust more, humans or AI.

Additionally, AI may reflect bias and lead to less effective treatment or lead to holding off treatment longer. For example, for years the kidney transplant waitlist mislabeled Black men as healthier than they really were, placing them lower on

---

[10] Cohen, 2019.

[11] US Attorney's Office. District of Vermont. Press release. (2020). Electronic Health Records Vendor to Pay Largest Criminal Fine in Vermont History and a Total of $145 Million To Resolve Criminal and Civil Investigations. https://www.justice.gov/usao-vt/pr/electronic-health-records-vendor-pay-largest-criminal-fine-vermont-history-and-total-145 (Practice Fusion is a criminal case; the EHR company took kickbacks from pharmaceutical companies in exchange for promoting their products; Purdue Pharma and other companies were involved in the scheme).

[12] US Attorney's Office. District of Vermont.

waitlists.[13] Data may continue to reflect the error. AI is described as a "ubiquitous, but often unseen, force in clinical decision making."[14] If healthcare consumers face discrimination, or if a recommendation relied on biased data, they may not know to ask whether AI was used. Knowledge about the use of AI may contribute to a healthy amount of skepticism and allow consumers to be one more check on a possibly skewed, biased recommendation.

While (hopefully) clinicians would obtain consent, if ambient software in a waiting room picked up facial expressions, body language, and words, and then a generative AI system wrote up a synopsis and added it to the healthcare consumer's EHR, that would violate the healthcare consumers' privacy, not meet expectations about the medical record, and could impact care going forward. Similarly, healthcare consumers would not know if waiting rooms equipped with games designed for the assessment of cognitive function have an AI system recording and analyzing actions or results.[15] Yet, when clinicians place AI systems without proper signs or disclosures, waiting for healthcare consumers to ask about such technology is unrealistic.

5.2.3 Automation Bias and Skepticism

There is noteworthy automation bias. Physicians are likely to defer to AI just as AI users in non-medical contexts tend to. Automation bias is common outside of health care: note the errors people following their GPS make or the confidence people have in rosy weather reports while actual rain is visible outside their window. Some have gone so far as to say that liquid must not be rain or there must be a road just past the cliff. Errors range from tragic to comical.[16] Blind trust is the root of following

[13] Auer, H. (contact) (2024) Friend or foe: A closer look at the role of health care algorithms in racial and ethnic disparities. Penn Medicine News. https://www.pennmedicine.org/news/news-releases/2024/march/the-role-of-health-care-algorithms-in-racial-bias-inequity.

[14] Penn Medicine News, 2024.

[15] Spoladore, D., Mondellini, M., Mahroo, A., et al. (2024). Smart waiting room: A systematic literature review and a proposal. Electronics, 13(2), 388. (describing uses of technology to enhance the waiting room experience; assessment of cognitive decline.)

[16] Mansharamani, V. (2020). Think for Yourself: Restoring Common Sense in the Age of Experts and Artificial Intelligence. Harvard Business Review Press.

bad advice. Automation bias has been noted in many areas including breast cancer,[17] criminal justice, and policing.[18]

Automation bias can harm healthcare consumers. The AI itself (due to data, algorithms, design, machine learning adjustments, and programming, etc.) and the incorrect use of AI both can cause errors. Physician and user errors include ignoring conflicting data or ignoring one's own knowledge gained from experience.[19] Automation bias may lead to a failure to question or corroborate the AI. Plus, time constraints also lead to pressure to move forward quickly rather than challenge AI recommendations.

Take for example a clinician who is somewhat skeptical of an anti-nausea drug. An AI popup alert leads the clinician to recommend the drug to a healthcare consumer despite his hesitation. Let's say research and the AI tool show that the anti-nausea drug decreases nausea slightly to moderately for 75% of people during the type of chemotherapy the person in question was having. However, the clinician never liked the side effect profile. If the consumer had known the clinician not only relied on the AI but ignored his own hesitation, the consumer may have been more likely to refuse the drug. Another relevant inquiry would be whether the consumer would be equally convinced if the 75% figure had come from the physician's experience and data collected at the physician's medical center as opposed to big data used in machine learning systems. Clinician automation bias increases the importance of the source of information. Automation bias gives a loudspeaker to AI and quiets the voice of the clinician's personal experience.

In one study, researchers compared the accuracy of physician decisions made with AI-assistance, without, and with the assistance of a biased AI tool. They found that with the biased tool, accuracy was lower than accuracy without an AI tool at all.[20] This demonstrates the ability of AI to cloud the physicians' judgment. "Even in controlled

---

[17] AI Bias May Impair Radiologist Accuracy on Mammogram. *Radiological Society of North America News.* https://www.rsna.org/news/2023/may/ai-bias-may-impair-accuracy ("Researchers from institutions in Germany and the Netherlands set out to determine how automation bias can affect radiologists at varying levels of experience when reading mammograms aided by an AI system;"..."“We anticipated that inaccurate AI predictions would influence the decisions made by radiologists in our study, particularly those with less experience," said study lead author Thomas Dratsch, MD, PhD, from the Institute of Diagnostic and Interventional Radiology, at University Hospital Cologne in Cologne, Germany. "Nonetheless, it was surprising to find that even highly experienced radiologists were adversely impacted by the AI system's judgments, albeit to a lesser extent than their less seasoned counterparts."”).

[18] O'Neil, C. (2016). *Weapons of Math Destruction.* Crown. (analyzing LA's PredPol.) Stephens, D. (2024). Forecasting justice: The promise of AI-enhanced law enforcement: While AI tools offer significant benefits in crime prevention and response, they also raise important ethical and operational concerns. *Police 1.* (noting automation bias.) https://www.police1.com/tech-pulse/forecasting-justice-the-promise-of-ai-enhanced-law-enforcement.

[19] Abdelwanis, M., Alarafati, H. K., Tammam, M. M. S., & Simsekler, M. C. E. (2024). Exploring the risks of automation bias in healthcare artificial intelligence applications: A Bowtie analysis. Journal of Safety Science and Resilience. https://doi.org/10.1016/j.jnlssr.2024.06.001.

[20] Jabbour, S., Fouhey, D., Shepard, S., et al. (2023). Measuring the Impact of AI in the Diagnosis of Hospitalized Patients: A Randomized Clinical Vignette Survey Study. JAMA. 330(23):2275–2284. https://doi.org/10.1001/jama.2023.22295

settings, without the usual pressures on time, clinicians favored automated decision-making systems, relying on the AI-based tool, despite the presence of contradictory or clinically nonsensical information."[21] Many studies support the conclusion that "physicians across expertise levels often fail to dismiss inaccurate advice generated by computerized systems."[22]

Automation bias is highly relevant to the argument for informed consent. Healthcare consumers should know that clinicians relied on AI so they can evaluate the recommendation. Healthcare consumers may not be aware of automation bias (unless they learn their lesson driving off cliffs or getting rained on). Without disclosures about the use of AI, they would be led to believe the doctor also used traditional diagnostic or treatment recommendation tools, manuals, expertise, and experience.

5.2.4 Right to Refuse and End of Life

Making an educated decision when refusing care is crucial as sometimes the informed refusal will allow a disease to flourish or bring life to an end. If AI improves the accuracy of prognosis, healthcare consumers may make decisions to forgo life sustaining interventions based on AI's predictions of low odds of survival. AI has the potential to impact end-of-life choices and concepts like a right to die as well as medical aid in dying. The right to refuse medical treatment is fundamental. Fifty years ago, in the *Matter of Karen Quinlan*, a young woman was in a persistent vegetative state and her parents wished to remove the respirator keeping her alive. A New Jersey court held that the state's interest in preserving life diminishes and right to privacy increases with poorer prognosis and a high degree of bodily invasion to continue life.[23] The *Cruzan* case further established some conditional right to direct one's own care even if refusing life sustaining care.[24] Nancy Cruzan was in a vegetative state and relied on a feeding tube for continued life. Cruzan was incapacitated, so her parents wished to make medical decisions on her behalf. They wanted to remove her feeding tube. The US Supreme Court held that the removal of life-sustaining care was permissible, but that a state may also set a high evidentiary standard for doing so. Missouri was permitted to require evidence that the removal of the feeding tube was what Nancy herself would have wanted. The Court noted that while states have an interest in the preservation of human life, individual liberty interests require states to recognize a competent person's right to refuse.

---

21 Khera, R., Simon, M., Ross, J. (2023). Automation Bias and Assistive AI: Risk of Harm From AI-Driven Clinical Decision Support. JAMA. 330(23):2255–2257. https://doi.org/10.1001/jama.2023.22557 (referring to Jabbour, et al., 2023.)

22 Kostick-Quenet, K., Gerke, S. (2022). AI in the hands of imperfect users. npj Digit. Med. 5, 197. https://doi.org/10.1038/s41746-022-00737-z.

23 In the Matter of Karen Quinlan, NJ Sup. Ct. 355 A.2d 647 (1976).

24 Cruzan v. Director, Missouri Dep't of Health, 497 U.S. 261 (1990).

Quinlan and *Cruzan* are both highly relevant to AI as AI tools used in clinical trials, diagnosis, prognosis, and treatment may make recommendations about the value or futility of life-sustaining care.[25] For example, AI tools projecting the likelihood of survival after resuscitation may lead consumers to request DNR orders and to request (by advanced directive or through a surrogate) ventilators be removed if they were placed. AI could improve the ability of cognitively impaired conscious people to communicate their wishes.[26] AI may add accuracy to determinations that people meet the criteria to qualify for medical aid in dying in states permitting it.[27] Similarly, hospice laws often require a prognosis of six months or less to live—AI's improved accuracy may allow people to choose non-curative comfort, and it may lead those predicted to live longer to re-think refusals of treatment. It is feasible that highly advanced AI systems will predict survival odds more accurately than humans do, leading to more educated decisions about whether to pursue aggressive disease treatment or refuse it.

Informed consent acts as veto power. Those not consenting deny the care (absent emergencies, etc.). Withholding consent to a clinical trial or intervention due to the role of AI in it is like any other refusal of medical treatment. It is permissible and may even be beneficial. For example, if a person knew that an AI tool was trained in the US where opioid deaths reached 100,000 per year in 2023, and the tool recommended opioids for pain, a refusal may be beneficial to the healthcare consumer, society, and the burdened healthcare system. Fraud in the design and use of the EHR system, as in the *Practice Fusion* case, can easily lead to imperfect information and prevent consumers from making an educated decision.[28] While a healthcare consumer could discuss an opioid recommendation with a doctor, automation bias or medicalization could lead the doctor to follow the AI's recommendation as well. In the case of opioids, clinicians should discuss other pain relief options. Especially in the psych industries, AI is likely to perpetuate medicalization as training data reflects a medicalized society. Decision support AI is not trained on data about people who use nonmedical solutions like diet and exercise without clinical oversight. AI may significantly contribute to clinicians recommending treatments that perpetuate a medicalized status quo. This should improve as data collected from wearables is

[25] Rose, S. and Shapiro, D. (2024) An Ethically Supported Framework for Determining Patient Notification and Informed Consent Practices When Using Artificial Intelligence in Health Care. CHEST, Volume 166, Issue 3, 572—578. https://journal.chestnet.org/article/S0012-3692%2824%2900633-0/fulltext (provides a hypothetical example of an AI tool recommending whether to change a code to "Do Not Resuscitate"; argues informed consent is crucial because the DNRAI tool is autonomous and alters the standard of care).

[26] Fins, J. and Shulman, K. Neuroethics, Covert Consciousness, and Disability Rights: What Happens When Artificial Intelligence Meets Cognitive Motor Dissociation? *Journal of Cognitive Neuroscience*. 2024; 36 (8): 1667–1674. https://doi.org/10.1162/jocn_a_02157.

[27] Battisti, D. (2025) AI-Assisted Suicide: A Game-Changer in Assisted Dying. PhilPapers.org. https://philpapers.org/rec/BATASA-5.

[28] US Attorney's Office, District of Vermont Press Release; US v. Practice Fusion, 2:20-cr-00011. District Court, Vermont. Deferred Prosecution Agreement. https://www.sec.gov/Archives/edgar/data/1124804/000156459020002291/mdrx-ex101_49.htm.

incorporated. Wearables may reflect the data of health-conscious people who track their own diet, fitness, and vitals.

The failure to disclose AI could contribute to consumers deciding to go along with treatments they would have refused. Studies show that physicians would refuse aggressive treatment themselves, but they rely on survival odds and data when conveying information to their patients.[29] The system seems to reward clinicians for recommending actions rather than for discussing the option of refusal or doing nothing near the end of life. AI tools may be designed to offer an array of interventions as options rather than suggest doing nothing. In such cases, clinicians may fail to adequately discuss the pros and cons as they see them. Such cases would be like the *Arato* case (discussed earlier) in which a failure to disclose the slim odds of survival led a healthcare consumer to consent to harsh chemotherapy. Conversely, if AI presented very low odds of survival based on a global data set, and a local hospital, perhaps a highly successful advanced cancer center, had significantly better results, a healthcare consumer risks refusing care based on insufficient information.

The World Health Organization (WHO) stated that patients refusing AI-enhanced care should not be denied treatment due to their refusal.[30] Ploug and Holm suggest that "the physician's role in the patient's formation of and acting on personal preferences and values" justify a right to withdraw from AI diagnostics and treatment.[31] Pruski suggests that in some cases it may become impossible to provide care to healthcare consumers opting out of AI in medical interventions.[32] If a healthcare consumer were to withhold consent to the use of AI, several problems and inconveniences could arise. In some cases, AI cannot be separated from treatment. In such cases, the AI either is the treatment or is a crucial part of it. Some AI-robotic surgeries may not have non-AI-robotic alternatives. Additionally, innovations leading people with paralysis to regain some movement do not replace older solutions to paralysis, as paralysis is considered incurable. AI may become the standard of care, and physicians may not be able to revert to an outdated technique or may not be comfortable doing so. A physician could face liability for using sub-par methods compared to the standard of care. Informed consent documents are not designed to allow sub-standard care.

The administrative burden of a healthcare consumer opting out of AI for their health records would also be high. Healthcare consumers should have the chance to opt out of sharing their data. To avoid data being used for medical research that goes against their values, ethically speaking, healthcare consumers should be able to opt out of broad consent provisions.

---

[29] White, T. (2014). Most physicians would forgo aggressive treatment for themselves at the end of life, study finds. *Stanford Medicine News Center*. https://med.stanford.edu/news/all-news/2014/05/most-physicians-would-forgo-aggressive-treatment-for-themselves-.html.

[30] Pruski, M. (2024). AI-Enhanced Healthcare: Not a new Paradigm for Informed Consent. *Journal of Bioethical Inquiry*. https://doi.org/10.1007/s11673-023-10320-0.

[31] Ploug, T., & Holm, S. (2020). The right to refuse diagnostics and treatment planning by artificial intelligence. *Medical Health Care Philosophy*, 23(1), 107–114. https://doi.org/10.1007/s11019-019-09912-8.

[32] Pruski, M., 2024.

5.2.5 Vulnerability of Healthcare Consumers

Sickness and injury may make people exceptionally vulnerable. There is already a significant and often noted power asymmetry between clinicians and healthcare consumers. The consumer may feel pressure to go along with aggressive care, experimental care, or a medical status quo that carries significant side effects. Healthcare consumers may be savvy; I do not want to underestimate them. Nonetheless sickness itself creates disparity, desperation, or even wishful thinking. Illness may cloud a person's judgment. It can do so in either direction—making them undergo risky interventions or refuse cumbersome but less risky ones.

Clinicians have many supports and advantages in their relationship with healthcare consumers. They have had medical education and work experience. The knowledge asymmetry is welcome—consumers expect clinicians to have expertise. The rigorous requirements of medical school, residency, board certifications, fellowships, and employers have led to a medical community in the US that is highly trained and knowledgeable. Sharing knowledge is a significant part of the informed consent process. As AI tools become even more ubiquitous, healthcare consumers should learn about their use. Yet to do so, consumers rely on clinicians and other healthcare professionals to disclose appropriate information.

5.2.6 Disclosure, Transparency, and Honesty as Promoting Trust

Trust in healthcare systems and institutions is complex. It tends to require trust in the humans involved. The public and healthcare consumers have two lines of trust at the point of receiving medical care: trust in the clinician and trust in the healthcare institution (hospital or hospital system).[33] AI may impact both and arguably brings another party and new trust relationships: that between the healthcare consumer and the tech developer and that between the clinician and the tech developer.

Honesty is a crucial precursor to trust.[34] Communication is a key element to building trust with healthcare consumers.[35] Disclosure, crucial to informed consent, is a communication to the consumer. When clinicians provide detailed explanations,

---

[33] Calnan, M., & Rowe, R. (2007). Trust and health care. Sociology Compass, 1(1), 283–308. https://doi.org/10.1111/j.1751-9020.2007.00007.x.

[34] Mission-Part, I. I. I. (2021). Honesty and Transparency, Indispensable to the Clinical. Business of Otolaryngology, An Issue of Otolaryngologic Clinics of North America, E-Book: Business of Otolaryngology, An Issue of Otolaryngologic Clinics of North America, E-Book, 55(1), 83.

[35] Greene, J., & Ramos, C. (2021). A mixed methods examination of health care provider behaviors that build patients' trust. Patient education and counseling, 104(5), 1222–1228. https://doi.org/10.1016/j.pec.2020.09.003.

they earn trust.[36] Nonetheless, deception remains common in health care.[37] Honesty and transparency are important ethical considerations at the root of informed consent. If clinicians are not forthcoming about the role of AI, the failure to disclose could impede trust and interfere with the doctor-patient relationship. Transparency about important aspects of care, although perhaps not about minutia, fosters trust.

Trust in relationships with clinicians and trust in the tools they use are different from trust in personal relationships. Trust in human relationships often does not require technical competence—humans may trust a spouse or friend despite their providing erroneous information or lacking skills. Trust involves both rational judgments and emotional factors like common identity and empathy.[38] Trust has been categorized as "thick" or "thin" to distinguish trust in a close family member from trust in a casual acquaintance.[39] A similar distinction applies to trust in an organization or sector, like health care. Trust in a clinician is widely varying. Some consumers develop strong personal ties with clinicians, sometimes in the course of a long-term critical illness. Even with an emotional attachment, health-care consumers putting their trust in clinicians should require clinicians using AI to have technological competence, a concept that should vary depending on the AI tool.

Trusting a chair not to break under one's weight is not about a "trustworthy" chair but speaks to trust in its designers and manufacturers.[40] Trust is perhaps better applied to people as people have agency and the ability to behave in trustworthy ways.[41] Some suggest that trust in the product generally equates to trust in the product's developers and in the researchers and clinicians using it.[42] In the context of AI, some frameworks and researchers refer to the trustworthiness of the product, which itself may be a proxy for trust in the people involved in its design and deployment. Accuracy, reliability, and generalizability (when relevant) speak to this issue. The EU Ethics Guidelines for Trustworthy AI are an indication that regulatory bodies are concerned with AI that may contribute to errors, bias, social harm, and other problems. The guidelines note three components of trustworthy AI: that it comply with

36 Greene and Ramos.

37 Hart, J. L. (2022). Deception, honesty, and professionalism: a persistent challenge in modern medicine. *Current Opinion in psychology*, *47*, 101434. https://doi.org/10.1016/j.copsyc.2022.101434.

38 Calnan and Rowe, 2007.

39 Calnan, M., & Rowe, R. (2007). Trust and health care. Sociology Compass, 1(1), 283–308. https://doi.org/10.1111/j.1751-9020.2007.00007.x.

40 Jones, C., Thornton, J., and Wyatt, J. (2023). Artificial intelligence and clinical decision support: clinicians' perspectives on trust, trustworthiness, and liability, *Medical Law Review*, Volume 31, Issue 4, Autumn 2023, Pages 501–520, https://doi.org/10.1093/medlaw/fwad013.

41 Jones, C., Thornton, J., and Wyatt, J. (2023). Artificial intelligence and clinical decision support: clinicians' perspectives on trust, trustworthiness, and liability, *Medical Law Review*, Volume 31, Issue 4, Autumn 2023, Pages 501–520, https://doi.org/10.1093/medlaw/fwad013.

42 Jones, et al., citing Baier, A. (2013) 'What is Trust' in Monique Deveaux and others (eds), Reading Onora O'Neill Taylor & Francis Group.

laws; that it be ethical; and that it be robust, both technically and socially.[43] Yet trust in AI is different from trust in humans—it is related to expectations. Humans can earn trust by proving themselves worthy. AI earns trust by consistently performing tasks well. Some suggest that the accuracy of the tool is more important than disclosures about AI tools.[44] "Computational reliabilism" justifies trust in medical AI without requiring explainability.[45] While clinicians may be quite comfortable with the computational justification, healthcare consumers would need to know that AI is being used so that they too can evaluate the AI's reliability. The disclosures and the informed consent process allow the consumer to become comfortable and trust the AI. The perception that AI is making mistakes impacts human trust in AI systems.[46] Communication about performance metrics has been shown to be a better way to develop trust than algorithmic explainability or transparency about how an algorithm works.[47] Put simply, an AI tool that consistently provides accurate actionable output may earn trust even if it is highly complex and humans cannot explain how it comes to its output. Conveying information about the accuracy and reliability of an AI tool would thus do wonders for promoting healthcare consumer trust in AI and the clinicians using it.

Trust in the clinician may be enough for some consumers, yet the decision to provide adequate disclosures should be based on objective standards. Clinician trust in an AI tool should not indicate that consumers have less need to know about an AI tool's use. Consumer trust in the clinician should not be used to justify skipping the step of informing the consumer. It would be paternalistic to suggest that because a hospital system purchased an AI tool and clinicians trust the tool, consumers do not need to know about it and sign off on its use.

Consumers may frame trust as dependability. Across health care, consumers should be able to depend on clinicians and the tools they use. Dependability accounts for the accuracy of an AI tool and speaks to whether consumers can rely on recommendations AI makes and clinicians using AI properly.

Additionally, public trust arguably would improve if there were a robust regulatory framework.[48] Public trust requires reliability, which may be more important to many consumers than technical explanations. "A relationship of reliance is based on reasonable expectations, proven capacity, open communication and clear and

---

[43] European Commission: Directorate-General for Communications Networks, Content and Technology and Grupa ekspertów wysokiego szczebla ds. sztucznej inteligencji, Ethics guidelines for trustworthy AI, Publications Office, 2019, https://data.europa.eu/doi/10.2759/346720.

[44] Pruski, 2024.

[45] Durán, J. M., & Jongsma, K. R. (2021). Who is afraid of black box algorithms? On the epistemological and ethical basis of trust in medical AI. *Journal of Medical Ethics*, 47(5), 329–335. https://jme.bmj.com/content/47/5/329.

[46] Zerilli, J., Bhatt, U., & Weller, A. (2022). How transparency modulates trust in artificial intelligence. *Patterns*, *3*(4). https://www.cell.com/patterns/fulltext/S2666-3899(22)00028-9.

[47] Zerilli, 2022.

[48] Kerasidou C., Kerasidou A, Buscher M., Wilkinson, S. (2022) Before and beyond trust: reliance in medical AI. *Journal of Medical Ethics*; 48:852–856. https://jme.bmj.com/content/48/11/852.

enforceable systems of responsibility and accountability."[49] The reasoning here is not just that the tech itself is reliable but that there is also a reliable system of oversight in place. Proper regulatory oversight would not relieve the need for informed consent, but it would contribute to consumer confidence.

There are many nuances and the relationship between trust in some cases is bidirectional—being informed fosters trust. Yet trust may inversely impact how people engage in medical decisions. For example, one study revealed that more trust may lead to less engagement. It suggests that people may trust doctors more when they are less informed themselves. Yet the same study found that among those with a high degree of trust, older healthcare consumers did not engage as actively in medical decision making as younger ones. Healthcare consumer preference in engaging with medical decision making was not dependent on trust, and in some cases the two were inversely related.[50] Informed consent allows consumers of care to make informed decisions. Yet it is not a guarantee of their interest or enthusiasm for controlling their healthcare journey—informed consent is merely a technicality for some consumers.

5.2.7 *Avoid Lawsuits for Violations of Informed Consent*

Rather than an ethics-based reason, avoidance of litigation is a practical reason to disclose the use of AI in medical care. Informed consent is a well-established right of healthcare consumers and does not negate their rights or equate to assumption of risk. Clinicians have well-known obligations to disclose material facts. The Colorado AI Act, California SB 3030, the Utah Artificial Intelligence Policy Act, and the Texas Responsible AI Governance Act impose disclosure obligations for generative AI .[51] Clinicians understanding the reasonable person standard should anticipate which situations call for informed consent by assessing materiality. Erring on the side of disclosure would help clinicians avoid breach of informed consent lawsuits. The framework presented here would help clinicians and consumers.

In addition to bringing lawsuits, to hold hospitals accountable, healthcare consumers can call on state or local health departments. Most states have an online portal where consumers can submit complaints. And the Department of Health and Human Services (HHS) evaluates complaints about violations in clinical trials. Hospital systems have an incentive to avoid informed consent violations. Getting in the habit of disclosure and sign-off would be helpful to both clinicians and consumers.

---

49 Kerasidou, et al., 2022.

50 Pokhilenko, I., van Esch, T., Brabers, A., & de Jong, J. (2021). Relationship between trust and patient involvement in medical decision-making: A cross-sectional study. PLoS One, 16(8), e0256698. https://doi.org/10.1371/journal.pone.0256698.

51 Colo. Rev. Stat. §6-1-1701, California Assembly Bill 3030, Utah Code Title 13 Chapter 2 Part 4, Texas House Bill 149.

5.2.8 *Data-Related Rationales*

The ethical use of data requires informed consent for many of the same reasons medical treatments or research require it. For example, it is reasonable to believe that many healthcare consumers would want to know how their data is being used; they may be unlikely to expect their data to be used in many of the ways in which hospitals use data; their decision to share data (and sometimes to move ahead with treatment) may be impacted by disclosures about how it would be stored and shared; a lack of informed consent may appear deceitful and pose a risk of legal liability; and disclosure is relatively easy. Broad disclosure has watered down consent; there should be an ethical duty not to share beyond the healthcare consumer's expectations without express consent even if healthcare systems must obtain a second signed informed consent document.

Consumers have a strong argument for informed consent when data is either repurposed for or collected for the purpose of training data for commercial AI medical tools. As described, both privacy and unjust enrichment justify a system of informed consent for the collection and use of data. Fairness is fundamental to consumer-based economies. The increase in consumer data privacy laws demonstrates legislative attention to concerns. Institutional policies can shore up consumer confidence by rigorous informed consent provisions. The potential for privacy lapses and cybersecurity breaches is high yet in many cases consumers of medical care must divulge personal and even intimate details to get appropriate treatment for diseases, disorders, and injuries.

5.3 TL;DR

Informed consent is a familiar tool in medicine. Clinicians and researchers understand its basic requirements. Consumers want to know when AI is used. AI's use may impact their decision making. Some healthcare consumers would not know to ask about it in many circumstances. They have no idea it might be used. And clinicians may defer to AI recommendations and overly rely on it due to automation bias. Consumers of healthcare may be vulnerable due to illness or socioeconomic positioning. Informed consent promotes trust; a failure to disclose the use of AI may be deceitful. A lack of informed consent could increase the risk of liability. Informed consent errs on the side of respect for the healthcare consumer. Common arguments against informed consent to the use of AI are discussed below.

Chapter 6
Arguments Against Requiring Informed Consent

6.1 Objections to Requiring Informed Consent

Some ethicists argue that informed consent is not necessary when it comes to AI in healthcare generally, except for some special situations.[1] For example, Pruski argues that its disclosure is ethically required in the case of certain surgical robotics and automated diagnoses, yet argues there is not an obligation to disclose the use of AI generally.[2] Others present it as the desired status quo but provide so many exceptions that the requirement to obtain informed consent becomes narrow—the exceptions swallow the rule. While if AI is used to turn on the lights or manage the parking garage, healthcare consumers need not be informed, depending on the degree of impact on medical decisions or data use, informed consent may be imperative.

In addition to the arguments here against requiring informed consent for the use of AI, bioethics literature has been increasingly critical of informed consent and even the role of autonomy. By portraying autonomy as "Western" and individualistic,[3] and informed consent as legalistic, scholars may be ushering in a new paradigm that undercuts healthcare consumers and human rights. Autonomy and informed consent arose to combat past abuses and to ensure the ability of healthcare consumers to

---

[1] Cohen, 2019.

[2] Pruski M. (2024). AI-Enhanced Healthcare: Not a new Paradigm for Informed Consent. *Journal of Bioethical Inquiry*, 21(3), 475–489. https://doi.org/10.1007/s11673-023-10320-0.

[3] For example, see Miola, José and Gilbar, Roy, One Size Fits All? On Patient Autonomy, Informed Consent and the Impact of Culture (October 10, 2014). University of Leicester School of Law Research Paper No. 14–23, Available at SSRN: https://ssrn.com/abstract=2508222; Milligan, E., & Jones, J. (2016). Rethinking Autonomy and Consent in Healthcare. Bioethics: Medical, Ethical and Legal Perspectives, 21. https://web.archive.org/web/20190501195910id_/https://cdn.intechopen.com/pdfs/53439.pdf; Hwang, S. S. I. (2025). An Eastern Principle of Relational Autonomy Shaped Attitudes Toward Mask Mandate During the COVID-19 Pandemic. *Voices in Bioethics*. 11. https://doi.org/10.52214/vib.v11i.13943; Spector-Bagdady, K., & London, A.J. (2025) Disclosure as Absolution in Medicine: Disentangling Autonomy from Beneficence and Justice in Artificial Intelligence. *The American Journal of Bioethics*, 25(3), 1–3. https://doi.org/10.1080/15265161.2025.2458424.

A. Zimmerman, *Informed Consent for AI and Health Care*,
SpringerBriefs in Ethics, https://doi.org/10.1007/978-3-032-11347-4_6

protect themselves and make decisions pertaining to their bodies, and later, their data. The turn against autonomy, the criticisms of informed consent as overvaluing the individual or suppressing the influence of the community, physicians, states, friends, and relatives is concerning. Community does shape individuals' worldviews and even their decision making—yet informed consent remains crucial to ensure individual rights. Globally, human rights are precious[4] and depend on states and institutions within states to ensure them. Some iterations of "relational autonomy" may prevent individuals from directing their own medical care, thus returning to a paternalistic system. The spirit behind the increasing attention to relational autonomy and the downplaying of the importance of autonomy as compared to beneficence and justice may somewhat set the tone of the bioethics literature about informed consent and AI in health care.

6.1.1 The Argument that Disclosure and Consent Forces Healthcare Consumers to Assume Risk

Joshua Hatherley asserts that disclosure may cause harm to healthcare consumers through its link to accountability avoidance.[5] He suggests that informed consent could reduce the accountability of various stakeholders, such as developers and professionals using AI. His argument assumes that informed consent is akin to assumption of risk. Yet physicians cannot use informed consent as a protection from medical malpractice. If they act with gross negligence, they may be susceptible to a valid claim. Disclosures protect from claims resting on the failure to disclose. Informed consent to the use of AI does not immunize most stakeholders in the AI ecosystem. It would not protect an AI developer from product liability. A healthcare consumer's signature also does not indicate the healthcare consumer assumes the risk of the hospital procuring poorly designed technology and using it or of an untrained physician using AI incorrectly, although medical malpractice requires the high standard of gross negligence.

When couched in the appropriate role of medicine and a duty of care that runs to healthcare consumers, informed consent is not an assumption of risk. John Kultgen describes the "notion that consent in any form waives the right to autonomy and justifies all interventions" as careless.[6]

Even when a person signs a waiver when engaging in an inherently dangerous activity, the person assumes reasonable risk but does not bear the risk of negligent or

---

[4] Raposo, V.L. (2019). Lost in 'Culturation': medical informed consent in China (from a Western perspective). *Med Health Care and Philos* 22, 17–30. https://doi.org/10.1007/s11019-018-9835-0 ("while still respecting Chinese particularities, this paper advocates that China adopt patient individual informed consent because this is the only solution compatible with human dignity and human rights.").

[5] Hatherley, J. (2024). Are clinicians ethically obligated to disclose their use of medical machine learning systems to patients? Journal of Medical Ethics. https://doi.org/10.1136/jme-2024-109905

[6] Kultgen, 1995.

intentional acts or omissions. For example, bungee jumping, zip lining, and skiing often require documentation releasing corporations or individuals from liability. In such cases, the duty to the risk-taking individual is altered or limited. Even having waived some ability to blame the company, the individual would be able to sue for reckless or intentional misconduct leading to harm.[7] In the case of informed consent, the duty of the clinician to the healthcare consumer is not altered by the consent documents.

Informed consent does not release the physician from the duty to provide services meeting the standard of care properly with due care. The argument that disclosures and consent shift the assumption of risk is weak. Medical malpractice claims often arise despite proper disclosures and informed consent. For example, consenting to surgery and knowing the risks would not diminish liability of a physician who leaves a surgical sponge or tool inside a healthcare consumer or one who operates on the wrong body part. Whether the context is AI or other medical tools, disclosure doesn't eliminate accountability.

Spector-Bagdady and London suggest "If health services are deficient ..., informed consent cannot—and should not—make their provision morally permissible."[8] And indeed it does not. Their point is that informed consent alone will not protect consumers enough. There is no reason to presuppose that informed consent would allow clinicians to use medical AI tools that are subpar, unreliable, inaccurate or otherwise inappropriate. Referring to Cohen suggesting the requirement for informed consent is strongest "where the physician lacks a good epistemic warrant to believe that the AI/ML recommendation is correct", the authors note that informed consent would not be appropriate until "safety, efficacy, and equity have been established."[9] It makes sense to limit the medical AI software and device menu and thus informed consent to AI tools that show promise, are proven, or are part of a carefully overseen clinical trial. The situation to which Cohen refers suggests a consumer wishes to follow the AI recommendation against the clinician's advice or when the clinician has no confidence in it yet relies on it anyway. While positioning that as a strong call for informed consent makes sense, it should additionally be seen as a strong call for better scrutiny at the point of purchase or deployment. Arguably, the FDA would not approve unreliable or inaccurate devices if they fall within the scope of regulation.

Informed consent does not expand the shopping cart to every unapproved or unreliable tool just through signing off—that is neither the practical use nor the spirit behind informed consent. (It is relevant to a right to try, to some aggressive human alterations, and the least proven interventions.) Tools about which the government is lax face the scrutiny of hospital procurement and clinicians before being adopted in

---

[7] Segal, (2013) Assumption of risk versus informed consent. Medical Justice. https://medicaljustice.com/blog/assumption-risk-versus-informed-consent/ (example of drunk bungee operator using a cord that is too long.)

[8] Spector-Bagdady, K., & London, A.J. (2025) Disclosure as Absolution in Medicine: Disentangling Autonomy from Beneficence and Justice in Artificial Intelligence. *The American Journal of Bioethics*, 25(3), 1–3. https://doi.org/10.1080/15265161.2025.2458424.

[9] Spector-Bagdady and London, 2025.

health care. Eliminating informed consent and leaving the use of AI to institutions and clinicians would be a return to paternalism. Those who criticize informed consent out of fear that informed consent would lead to low-quality AI tools should be cautioned that a lack of informed consent would be even less protective of consumers. Moving the only approval of the use of AI to the institutional level would undermine consumers. Informed consent to the use of an unproven, experimental product outside of a clinical trial would be better couched in a "right to try." As many bioethicists note, informed consent is not the only consideration. Yet its insufficiency is not a reason to cast it aside.

Additionally, the standard of care operates as a key ethical component of the delivery of health care and is important in medical malpractice lawsuits. Healthcare consumers have a better argument that they have been wronged when they are denied the standard of care when meeting such a standard would have been possible. If an AI system is the standard of care, and its use is disclosed to the consumer and the consumer consents to its use, for example, in the context of surgery or medical treatment, a clinician still must use it properly.

As far as privacy and cybersecurity, hospital systems are held to high standards and consenting to data collection, storage, and use is not a waiver of rights if proper data protections are not in place. Consumers would not assume cybersecurity and privacy risks specifically although once a breach occurs, they may succumb to some harm.

6.1.2 The "Just a Tool" Argument

The argument suggests that AI is a tool like many other tools a doctor might use, including books for research, a scalpel, or an x-ray machine.[10] This argument relies on the following premises. First, it assumes that professionals use AI in a way that is not ethically different from how they use other tools; put differently, it assumes the AI performs tasks that are ethically like the tasks of those other tools. Second, in some cases, it assumes that AI makes a mere suggestion or assists physicians performing a task rather than performing it for them or actively diagnosing or recommending treatment. Neither premise holds true when challenged. AI is categorically unlike other tools in that it makes recommendations, gives choices, categorizes, and even generates content. It may assist or perform a task that the practitioner used to do personally. For example, if machine learning generates meeting notes, there may be bias, incorrect information, or a perspective that ignores some relevant context. While it may be less biased than a clinician, it remains unlike non-AI assistive tools.

---

[10] Park, H.J. (2024). Patient perspectives on informed consent for medical AI: A web-based experiment. Digital health, 10, 20, 552, 076, 241, 247, 938 (finding that patients perceive AI as different from consulting a colleague; research findings support AI not being "just a tool".) https://doi.org/10.1177/20552076241247938; Cohen, 2019, What to Tell the Patient; Hatherley, 2024.

Healthcare consumers weigh information provided by physicians who disclose using AI differently from information based on physician expertise or on consultation with another physician.[11] Knowing that information is perceived in varying ways based on its source should lead one to conclude that to properly analyze information, healthcare consumers need to know its source. Anti-AI bias differs from bias in favor of the non-AI assisted judgment of clinicians. A desire to please the physician may be stronger than a desire to follow the recommendations AI provides. Or the opposite may be true for those with more confidence in AI than in the doctor's expertise.

Additionally, because books, journals, and their online versions have been around for a long time, it is highly likely healthcare consumers assume physicians learn from books and journal articles rather than from AI, which continues to generate erroneous summaries. Many would agree that clinicians should not rely on AI overviews provided at the top of Google search results rather than the underlying articles. With the current state of some generative AI, it would not be appropriate to presume consent. Presumed consent to the use of medical books is noncontroversial, whereas we would not presume consent to using the Bible, social media, or commercial fiction as a medical source. AI's medical information may be largely correct, but even the most reliable tools represent a departure from the long-term status quo.

Technology is generally not neutral in the way some other tools may be. Certainly, one could argue books, lectures, and scalpels are also not neutral, and that they are used in ways that reflect values. When the concept of neutrality is applied to AI, it is a bit like the well-known and rather flawed "guns don't kill people; people kill people" argument. While the intentional or accidental inappropriate or harmful use of AI is a possible problem, the user has less control over AI than over other tools. Its seemingly appropriate use could perpetuate bias, lead to discrimination, create errors, and lead to poor medical recommendations. The use of a scalpel to commit a crime or to violate informed consent by operating on the wrong body part or person is different from the use of AI. The user has more control over non-AI instruments. With AI, clinicians lose some control; the clinician trying to do everything right could still succumb to algorithmic bias or a misdiagnosis due to overfitting inherent in the AI or misunderstanding the limitations of the AI tool. Other tools have known limitations, but generally they cannot go rogue and their influence over physicians is transparent and limited.

AI is expected to become increasingly "agentic", meaning new tools will be more proactive, more able to contextualize, and more able to carry out tasks autonomously. For example, autonomous cars have AI mimicking human awareness of many road conditions, directions, and obstacles. Medical AI that performs tasks from start to finish without following specifically programmed steps has an outsized opportunity to diagnose, recommend treatment, carry out surgeries, and prescribe. Agentic AI may

---

[11] Findley, J., Woods, A., Robertson, C., & Slepian, M. (2020). Keeping the patient at the center of machine learning in healthcare. The American Journal of Bioethics, 20(11), 54–56. https://doi.org/10.1080/15265161.2020.18201008

understand the "goals and visions"[12] of the user and carry them out, leaving humans out of the organization of logical steps that have always been part of medical care. Healthcare consumers may have a stronger right to know about it due to its fundamental differences from other tools and its users having less control over it. While it is unlikely that researchers, clinicians, and bioethicists would argue against disclosing its use in treating consumers, they may debate disclosing agentic AI in managerial tasks and customer communication. For example, AI tools perform tasks in assisted living and in pre- and post-op coordination.[13] Informed consent should be included with medication reminders and treatment-related aspects as well as communications. As AI tools perform both nonmedical and medical tasks in one tool, it would be best to obtain informed consent even if some of the tasks are outside the scope of health care.

Encased in the "just a tool" argument is the argument that AI will be completely ubiquitous and using it is like using electricity,[14] turning on the lights, heating the office, and doing other everyday tasks. This line of reasoning is based on both its pervasiveness and its ability to ease or perform many everyday tasks. However, such comparisons call for critique. While electricity was a disruptor and altered family and work life considerably, it did not recommend, classify, categorize, or create content. Early electricity replaced oil lamps and lanterns; cars replaced horse-and-buggies. But AI competes with human tasks in a different way. AI performs many tasks better than humans and even performs ones that humans, even those with very high intellect, would not be able to. And many types and uses of AI compete with human work, thought, and creativity. Some tasks that AI could be trained to perform are at the center of what makes humans special or human creative work personal, heartfelt, and unique. Furthermore, in health care, AI can perform tasks and even be trained to make judgment calls previously reserved for trained clinicians. While the language framing the issue is open for debate—was electricity a more powerful disruptor?—electricity and other background inventions are categorically unlike AI. The risk of AI integrated in a wearable or surgical robotic, or prescribing drugs (as a pending bill in Congress would allow)[15] is simply a more complex risk than turning on the lights. Ubiquity should bolster calls for informed consent, not lead to shrugging off its importance.

---

[12] Purdy, M. (2024) What Is Agentic AI, and How Will It Change Work? Harvard Business Review. https://hbr.org/2024/12/what-is-agentic-ai-and-how-will-it-change-work.

[13] Purdy, 2024 ("…Sarah, an AI agent who "radiates warmth and understanding" while providing help with assisted living. Sarah can ask patients about their day, organize menus and transport, and regularly remind patients to take their medication. Judy, another AI-powered agent, helps patients with pre-operative procedures, for example by reminding patients about arrival time and locations, or advising on pre-op fasting or stopping medications."

[14] Douetteau, F. (2022). Everyday AI could become as ubiquitous and necessary as electricity. *Venture Beat.* https://venturebeat.com/ai/everyday-ai-could-become-as-ubiquitous-and-necessary-as-electricity/

[15] H.R.238 - Healthy Technology Act of 2025. United States Congress. https://www.congress.gov/bill/119th-congress/house-bill/238/all-info (would allow machine learning tools to act as practitioners able to prescribe if the state approves and the tool is FDA approved.)

6.1.3 The "AI is Merely for Decision Support" Argument

Some argue that AI decision support systems need not be disclosed because they recommend rather than make clinical decisions. However, automation bias, complacency, hesitation to challenge AI's recommendations, and even clinical time constraints due to seeing many healthcare consumers for short appointments lead clinicians to defer to the AI tools' recommendations instead of using their own clinical judgment. Due to just "going along" with the AI, the AI recommendation may be outcome-determinative rather than merely influential. And healthcare consumers are likely to believe that clinical expertise, context, experience, and data led to recommendations that clinicians relay to them. Many healthcare consumers do not know that AI is used this way. Furthermore, automation bias itself can lead clinicians to overlook non-automated considerations, including context, and ignore conflicting data from other sources. AI could err or prevent the clinician from exploring broader options. Therefore, clinicians should disclose the use of decision support systems.

6.1.4 The "Frees up the Doctor's Time" Argument

AI improves administrative efficiency. Clinicians are often overworked, fatigued, and distracted from clinical care due to paperwork. Yet the benefit of alleviating administrative burdens does not justify failing to disclose its use and obtain informed consent when the AI in question is material to consumer decision making. When comparing the pros and cons of any AI, certainly administrative efficiency is a pro. There are also risks to some AI tools that save time. They may be biased, lead to discrimination, create a patient story that is not an accurate or fair reflection, and create a larger set of data vulnerable to cybersecurity breaches or accidental viewing. If generative AI summarizes healthcare consumer charts, appointment notes, or behaviors, healthcare consumers should be aware and have the chance to review any such summaries.

Some argue that the increased efficiency may be a reason to not disclose AI's use or seek consent. Yet, administrative efficiency outweighs the lost efficiency of obtaining informed consent. Administrative tasks are the most common use of AI in medicine. People generally accept that EHR systems store a significant amount of personal information. Because the information can be highly prejudicial and stigmatizing, healthcare consumers should have an opportunity to provide consent or to opt out of data uses. While it may be impractical to opt out of administrative AI in EHR systems, informed consent should apply to clinical decision support and generative AI tools embedded in the systems as well as to data collection. It would be difficult to have a healthcare consumer request their entire medical history be held using only paper records, i.e., to opt out of the EHR system altogether. Communicating with payors would be especially slow and unworkable. AI solely for administrative efficiency that does not touch on medical or personal data, clinical care, or research is unlikely to justify the burden of informed consent. It will be increasingly important for clinicians to understand the data use landscape of EHR tools.

Clinical decision support tools may free up clinicians' time. Some worry that they will eventually replace clinicians altogether—thus acting as medical devices and offering medical advice. Clearly, healthcare consumers will benefit from the administrative and clinical efficiencies AI brings whether they prefer AI or a clinician. However, the efficiencies do not impact the ethical impetus for informed consent. This view of informed consent as a barrier to efficiency makes assumptions that people would opt out if given the choice. Informed consent is not a barrier; it is an opportunity to foster a trusting relationship and to knowingly participate in the results of innovation.

6.1.5 The "Transparency is Enough" Argument

Informed consent requires signoff and conveys that the healthcare consumer has some understanding. Some argue that it is not needed or is redundant. The concept of a right to notice and explanation in the Blueprint for an AI Bill of Rights[16] is that the public should be aware of AI and able to obtain information about its use.[17] Applied to healthcare, the AI Bill of Rights notice and explanation rights could look like a lesser requirement, not a part of informed consent. There is an argument that suggests that with transparency, clinicians may presume or assume consent. But transparency is not enough in many healthcare contexts for several reasons. When disclosures meeting the principle of transparency are posted on a sign or website or mixed into fine print, they are often overlooked. Transparency has no requirement that the reader read the information let alone approve it. It is one directional. In some settings, especially when access to websites is contingent on some notice, tech users check boxes and click through cursory banners asking them to allow cookies so often that the motion becomes rote. The signoff is merely a way to get the question of consent off the screen and move on. Informed consent in the healthcare setting is more than transparency and is a meaningful way to ensure that healthcare consumers understand and wish to proceed with clinical care, research, or providing data, etc. It is more than notice and explanation.

---

[16] US. Office of Science and Technology Policy's (OSTP).

[17] Hurley, M. E., Lang, B. H., Kostick-Quenet, K. M., Smith, J. N., & Blumenthal-Barby, J. (2024). Patient Consent and The Right to Notice and Explanation of AI Systems Used in Health Care. *The American Journal of Bioethics*, *25*(3), 102–114. https://doi.org/10.1080/15265161.2024.2399828.

6.1.6 The "Healthcare Consumers Would Ask if They Wanted to Know" Argument

As noted earlier, some consumers of medical care would not know to ask. Even with mainstream media and scientific coverage of AI in health care, it is not right to leave asking about AI up to the consumer.

6.1.7 The "Stifles Innovation or Leads People to Deny Care" Argument

This argument targets disclosure and the potentially chilling effect that disclosing the use of AI could have on healthcare consumer participation in medical care. This argument assumes that consumers would walk away from the benefits of AI in medical care. There are two important objections: first, when disclosures lead people to refuse care, that may be a good check on medicalization, overprescribing, and over-treating. If healthcare consumers are avoiding appropriate medical care because of AI, then information about the benefits and the role of AI should be improved. Denying care is framed as bad, like the word "noncompliant" for one who intentionally does not take a prescribed drug. Second, disclosure should not have a chilling effect on innovation because many people want AI to cure tricky illnesses or because they feel AI is a significant advance that will improve their care. While the information should not be like that which a sales and marketing team would use, it also need not be conveyed as a warning or a negative. Many consumers of medical care would welcome innovation in complex cases.

Institutions and government agencies note the importance of data to innovation. Many healthcare consumers share their data in hopes of innovative approaches to chronic and acute diseases and disorders. Informed consent would not stop them from sharing; it would clarify which uses of data they consent to.

A pandemic study which focused on perceptions of whether AI or clinicians better recognized healthcare consumer's unique characteristics suggests that patients who are less trusting of doctors are more open to medical AI.[18] Some healthcare consumers would happily consent and prefer AI. Failing to disclose its use to them would not

---

[18] Frank D-A, Elbæk CT, Børsting CK, Mitkidis P, Otterbring T, Borau S (2021) Drivers and social implications of Artificial Intelligence adoption in healthcare during the COVID-19 pandemic. PLoS ONE 16(11): e0259928. https://

doi.org/10.1371/journal.pone.0259928; National Academy of Medicine. 2025. *An Artificial Intelligence Code of Conduct for Health and Medicine: Essential Guidance for Aligned Action.* Washington, DC: The National Academies Press. https://doi.org/10.17226/29087 (regarding AI assisted documentation, "However, the use of such AI tools raises concerns that patients may not understand how AI is being used in their care, potentially compromising their autonomy. Even if patients have the option to consent to—or decline—the use of AI, they may not fully comprehend what they are consenting to or declining.").

serve to incorporate AI seamlessly in health care. If AI is a draw to some people, transparency about its use would attract them to medicine.

The argument that informed consent would stifle innovation or lead to consumer refusals also has a paternalistic theme. At worst, it frames consumers of care as unwilling research subjects who would go along only if deceived instead of willing and interested consumers. It has an underlying tone of deception, as if consumers of medical care are more likely to go along the less they know.

6.1.8 *The Argument that Healthcare Consumers Will not "Understand the AI"*

Many suggest that people may be confused by AI.[19] This argument hinges on the idea that AI is too complex for consumers to grasp. Some even suggest that those who cannot ascertain a high-level understanding of an algorithm would then be ill-equipped to authorize its use, and thus deprived of its benefits.[20] While some use this premise to suggest more transparency and more detailed disclosures, others use it to suggest that explaining how AI is used and how it works is not fruitful. The argument tends to suggest that less information is better when more information would confuse healthcare consumers, create anxiety, or be beyond their scope of comprehension due to their educational level. These arguments not only belittle healthcare consumers, but they also lead to treating all healthcare consumers like the lowest common denominator—the ones least able to absorb the information and make educated decisions.

This type of argument also correctly acknowledges the complexity of opaque systems which cannot be explained—black box systems.[21] While in many cases, such tools will not make it through approval hurdles or institutional purchasing, many such tools hold varying degrees of promise in medicine. While notice and explanation implies explanation of how an algorithm comes to its output, an explanation of what is knowable could suffice if the AI tool is proven accurate and beneficial. Clinicians must be transparent about what they do not know as well as what they know about how a tool arrives at an output. There is no need to disclose every aspect of engineering in an AI tool if the tool is unexplainable or trade secrets prevent clinicians from knowing certain details. Primarily, to meet the requirements of informed consent, clinicians should disclose that the AI is being used, for what purpose, and describe the model based on its design or architecture. Healthcare consumers will get the gist, and once they have the basics, they may have many follow-up questions. Importantly,

[19] Hurley, et al., 2024 https://doi.org/10.1080/15265161.2024.2399828 (noting some will not understand due to the complexity of the AI, the limitations and communication styles of clinician, and the consumer's health and AI literacy, but suggesting that a more meaningful concept may be an understanding of the risks rather than an understanding of how the AI works.)

[20] Hurley, et. al., 2024.

[21] Hurley, et al., 2024.

some suggest that clinicians need significant reskilling to learn about AI themselves to convey accurate information to consumers of medical care. It is highly likely that many clinicians will learn more about AI used in health care as they become experts at using AI tools. However, many will not become engineers and do their own coding. Clinicians are usually users of AI tools rather than developers. As such, they too need to learn the basic model architecture, some information about the training data, the algorithm (if available), and the limitations. For explainable AI, the details of the code, or algorithm, may be available but may go beyond the scope of what is required for informed consent. Generally, clinicians will likely focus on learning to effectively use AI tools and developing awareness of their limitations. Some of the relevant information may come from the developer or the institution's IT, engineering, bioinformatics, or other departments. It can be relayed to healthcare consumers so that they can make informed decisions.

6.1.9 The Argument that the Administrative Burdens of Opting Out of AI or Seeking Informed Consent Are Too High

For administrative AI built into EHR systems, this may be true. However, the administrative burden of opting out of some medical AI could be vastly reduced as EHR system design anticipates consumers will wish to opt out. For example, generative AI for note summarizing and AI tools for recommending prescription drugs could be designed as options that clinicians can use but are not bound to. In many cases, tools offer these options already. When clinicians plan to use these features, informed consent allows consumers to object to them. Yet informed consent is not just about deciding among options. Disclosure of risk and understanding AI tools is especially important when choices are limited, or even singular.

When it is impossible to opt out, informed consent may still be valuable. It promotes transparency and allows consumers to understand AI's role in their treatment, diagnosis, prognosis, data collection and use, etc. Informed consent's sign off should not be just for show. It should be meaningful. By allowing consumers to ask more questions, clinicians may realize they need innovative ways to allow them to opt out if opting in violates their morals or autonomy. As noted throughout, WHO and other agencies hold that consumers opting out of AI should still be able to opt in to care without it.

AI is decreasing the administrative burden of informed consent. It is easier than ever to create clear explanations of medical and administrative AI tools and provide informed consent documentation. Some suggest an overload of provisions subject to informed consent undermines consumer attention to each provision.[22] Yet the use of AI is important to medical decision making and consumers deserve its use to be disclosed.

---

[22] Spector-Bagdady and London, 2025.

6.2 TL;DR

Those arguing against informed consent present several lines of reasoning including that informed consent forces consumers to assume risk, that AI is just a tool like many others, that AI supports but does not make decisions, that AI frees up the clinician's time, that transparency is enough, that healthcare consumers would ask about it if they wished to know, that informed consent could stifle innovation, that the healthcare consumer would not understand the AI, and that it is too difficult to allow consumers to opt out.

Chapter 7
Framework for Disclosure

7.1 Comparing Frameworks and Concepts

Many frameworks consider informed consent within the evaluation of healthcare AI tools. For example, in the recently published FAIR-AI framework, consent is couched in vulnerability, transparency, and equity best practices:

> There is an ethical obligation to notify patients when AI is being used and, when appropriate, to obtain their consent—particularly in sensitive or high-stakes situations. This obligation is heightened when there is no human oversight, when the technology is experimental, or when the use of AI is not readily apparent.[1]

The FAIR-AI framework incorporates consent into the process of categorizing AI tools or uses as low, moderate, or high risk. A lack of an informed consent process in the established "workflow" may contribute to risk, calling for more scrutiny, and possibly an ethical obligation to require it.[2]

Some frameworks, like FAIR-AI, and articles interpreting notice and explanation[3] obligations, apply a sliding scale determining whether some type of agreement that is less than informed consent is permissible and even preferable. For example, suggestions include providing a verbal, but not written disclosure, not requiring a signature (allowing a mere nod of agreement), or even presuming consent to AI upon written consent to the underlying medical treatment. The approach here differs from that – and recommends erring on the side of informed consent, with variance in the type of disclosure and the amount of detail required. If a sliding scale were to apply to the informed consent process itself, the result would be that in many situations where AI

---

[1] Wells, B. J., Nguyen, H. M., McWilliams, A., et al., FAIR-AI Consortium (2025). A practical framework for appropriate implementation and review of artificial intelligence (FAIR-AI) in healthcare. *NPJ digital medicine*, 8(1), 514. https://doi.org/10.1038/s41746-025-01900-y

[2] Wells, et al., 2025.

[3] Hurley, M., Lang, B., Kostick-Quenet K., Smith J., and Blumenthal-Barby, J. (2025). Patient Consent and the Right to Notice and Explanation of AI Systems Used in Health Care. *Am J Bioeth.* 25(3):102-114. https://pmc.ncbi.nlm.nih.gov/articles/PMC12143229/

A. Zimmerman, *Informed Consent for AI and Health Care*,
SpringerBriefs in Ethics, https://doi.org/10.1007/978-3-032-11347-4_7

is considered low risk, the healthcare consumer could remain completely oblivious to its use. Depending on the AI tool, I am hesitant to endorse the suggestion that "background" uses of AI be kept from healthcare consumers or be shared in notices rather than consent documents as they may be material. When AI has potential to be material to decision making, the open issue would be what to disclose.

One framework that Mello, et al. apply to tools that require human oversight is dismissive of informed consent and provides rationales for its lax approach for even high-risk AI tools. The authors suggest that AI is de facto not material if the tool cannot be avoided in the course of health care.[4] They describe categories which should not require notice to consumers at all. Nonetheless, the authors suggest that hospital systems should have an available AI policy and should allow healthcare consumers input in governance. While this is similar to those suggesting regulations should be stricter or that some high-level policies obviate the need for informed consent,[5] I suggest informed consent is an additional check on the hospital system. Governance that is friendly to healthcare consumers does not change ethical obligations to attain informed consent to many, especially risky, uses of AI.

Rose and Shapiro recommend a framework that includes whether the AI is patient-facing as a criterion for applying informed consent to its use.[6] While this is helpful, AI often works behind the scenes yet greatly impacts consumers. The role of informed consent requires an identifiable consumer of medical care, but it need not be limited to direct contact with the consumer. Behind the scenes, AI may be impacting clinical care and healthcare consumer outcomes. Rose and Shapiro's framework presents criteria and a scoring system for requiring informed consent. They include the degree of autonomy the AI tool has, deviation from the standard of care, whether the AI is patient facing, and the clinical risk. After assessing those, they also consider the administrative burden of informed consent.[7] Their approach is clean and easy to apply. It emphasizes the need for informed consent when AI tools are the most autonomous and allows for notice to suffice for tools that assist physicians. It differs from the framework here in two important ways: Rose and Shapiro's framework does not address the tendency to rely on AI when it is not meant to operate autonomously; and, while deviation from the standard of care is a good justification for informed consent, AI as the standard of care also justifies informed consent.

Generally, there is a tendency to equate the standard of care with less risk in the healthcare context. For example, if surgery is the standard of care for appendicitis, then a healthcare consumer would likely expect surgery when presenting with a case of appendicitis. However, it is becoming increasingly common to treat appendicitis with antibiotics. In some cases, antibiotics may pose more risk in that if they fail,

[4] Mello, M., Char, D., Xu, S. (2025). Ethical Obligations to Inform Patients About Use of AI Tools. *JAMA*. 334(9):767–770. https://doi.org/10.1001/jama.2025.11417

[5] Spector-Bagdady, K., & London, A.J., 2025.

[6] Rose, S. and Shapiro, D. (2024) An Ethically Supported Framework for Determining Patient Notification and Informed Consent Practices When Using Artificial Intelligence in Health Care. CHEST, Volume 166, Issue 3, 572–578. https://journal.chestnet.org/article/S0012-3692%2824%2900633-0/fulltext

[7] Rose and Shapiro, 2024.

the appendix could burst. However, surgery also poses many risks. As AI becomes more prevalent in medicine, its use is likely to become the standard of care in some cases. For example, if AI radiology tools are better at classifying than radiologists are, they may become commonplace. They are widely used already. A hospital may erroneously feel less need to disclose that which is standard. In arguing against the requirement of informed consent, or that informed consent is "illusory", Pilkington and Binkley state "one reason is that AI-clinician collaborations may be superior to clinicians acting alone and may become the standard of care."[8] However, superiority has little to do with informed consent. Informed consent is not a tool by which consumers agree to either sub-standard or non-standard treatment. *Becoming the standard of care should not eliminate the need for disclosure and informed consent.* In routine surgeries, clinicians disclose a significant amount of information about anesthesia, the surgical procedure, the anticipated recovery time, need for physical therapy, and risk of infection and seek informed consent from the healthcare consumer. If AI were involved somehow—for example, through a surgical robotic—a reasonable consumer of the surgery would likely expect to be informed about the AI. The standard of care is highly relevant to consumers in that if use of AI is the standard of care, then consumers of care may feel entitled to it. Nonetheless, the Rose and Shapiro framework may be beneficial when evaluating experimental AI, which is likely to be riskier than AI that has survived scrutiny, stood the test of time, and become the standard of care. The more uncertainty, the stronger the argument to disclose AI's use.

Some suggest that disclosure of how AI is used in care matters more than a description of the AI tool's basic inner workings.[9] For example, to Pilkington and Binkley, "The mechanics of why and how the generative chat is produced or clinical prediction is made is less important than how the clinician will use that information in caring for the patient."[10] They use an analogy to what healthcare consumers generally would want to know about a research study process or number of participants. However, in cases in which the risks are higher, it would seem very feasible that a reasonable person would want to understand, even at a broad and basic level, the AI tool. In some cases, how the AI tool generally works (its architecture, dataset, or conceptual information about its algorithm or design) is as important to consumers as understanding its role in their care. Not all disclosures should be reduced to "AI will be used in your care for xxx purpose."

[8] Pilkington, B. and Binkley, C. (2025) Clarifying When Consent Might Be Illusory in Notice and Explanation Rights. The American Journal of Bioethics. 25:3, 126–128.

[9] Mello, M., Char, D., Xu, S. (2025). Ethical Obligations to Inform Patients About Use of AI Tools. JAMA. 334(9):767–770. https://doi.org/10.1001/jama.2025.11417

[10] Pilkington, B. and Binkley, C. (2025) Clarifying When Consent Might Be Illusory in Notice and Explanation Rights. The American Journal of Bioethics. 25:3, 126–128. https://doi.org/10.1080/15265161.2025.2457728

7.2 Risk Appetite

Clinicians and consumers may vary greatly when it comes to uncertainty. One approach to uncertainty is analyzing risk appetite. Risk appetite varies personally. Some healthcare consumers choose risky interventions like experimental treatments while others tend to play it safe. Individuals vary from a life-at-all-cost priority to an aversion to cumbersome side effects. Furthermore, the consumer's curiosity may be satisfied by the availability of public information on the drug and by its widespread use. It is difficult for institutions to assess consumer risk appetite in deciding which information must be disclosed. The reasonable person standard allows clinicians to standardize a minimum baseline for what they disclose without a need to predict or assess which information a particular consumer will find meaningful to their decision making. Erring on the side of disclosure is preferable, ethically speaking.

Institutions also have varying appetite for ethics-related risk. Ethics failures ruin reputations. Hospitals are also clearly concerned about informed consent infractions for which they may be held liable. While the reasons differ, both parties to the informed consent documents have an interest in the consumer of care being well informed.

7.3 Applying a Risk and Reliance-Based Framework

The need to obtain informed consent and the level of disclosure itself depend on the type and degree of risk that AI poses,[11] as well as the potential for automation bias to lead clinicians to overly rely on the AI. Informed consent is the gold standard. Thus, the focus here is on how to determine what to disclose to enable consumers to make educated decisions about AI tools in their health care. As with non-AI traditional medical disclosures, both remote risk of catastrophic outcomes and high risks of common, but less severe, side effects should be disclosed. And like with non-AI medical disclosures, the uncertain areas in between will be most controversial. Increased risk calls for more detailed disclosures. The relationship between risk and the need for consent is well-established.

The simple framework I recommend has two steps: first, evaluate the likelihood and the degree of harm; and second, assess risk that a clinician would overly rely on AI. Rather than answering all relevant questions, the framework is an analytical tool.

[11] Kiener, M. (2021). Artificial intelligence in medicine and the disclosure of risks. AI & Society, 36(3), 705-713. https://doi.org/10.1007/s00146-020-01085-w (explores cybersecurity risk, bias and discrimination, and mismatch.); Findley, et al. 2020 (risk of violating autonomy.)

7.3.1 Step 1: Risk

The types of risk associated with the use of AI vary:

1) Data related: privacy, data sales and unjust enrichment, surveillance, psych-related, reidentification, reputational damage or incrimination, exploitation, financial, control over records and generated notes/stories.
2) Physical: injury to body/organs, death, forgoing appropriate treatment/disease progression, surgical error, and dosing error.

There is not a fine line dividing risks — many AI tools pose multiple ethics-related risks. For example, the EHR system includes decision support software, which may recommend drugs and support diagnosis, but also may use generative AI for appointment notes or ambient AI to assess mood, statements, or appearance.

There is not an agreed upon benchmark for measuring risk. Comparing the risk the AI poses to the risks a human performing the action poses is a strategy for some situations but is not always useful. For example, diagnosis by observing symptoms carries some risk of human error. Using AI may carry less risk of medical error generally but pose risks in special cases or when applied, for example, harms due to demographic variability or training data homogeneity. Bioinformatics AI tools perform computations bridging math and statistics with biological data. In complex bioinformatics, a human would not be able to perform some of the computations, nor navigate big data in a timely fashion. Therefore, the comparison dissolves. Comparison to the non-AI (or older AI) standard of care is not always the best way to assess risk.

As noted, another way of measuring risk could be distance from the prevailing standard of care.[12] However, AI may be less risky than the standard of care in many tricky, intractable, or incurable diseases. In such cases it may become the standard of care. While it has not yet been a panacea, overstating the risk would not be appropriate. AI should reduce the risk of clinical errors significantly and improve upon clinical decisions. For example, using AI to determine the likelihood that a cancerous tumor will be sensitive to a drug reduces the risk that a person would bear harsh side effects with a lower likelihood of success. As noted, even the standard of care requires careful informed consent and significant disclosures, so straying from it may not be as highly relevant as some suggest.

Within each of the two broad categories, data-related and physical, there are degrees of harm. Clearly death, organ damage, and the like are significant concerns, the risk of which calls for detailed disclosure. A minor short-lasting physical discomfort could require less detailed disclosure. A breach of identifiable data of a consumer who has a plethora of disorders that tend to be stigmatized could be a threat to the consumer's reputation and livelihood. That would be worse than a rogue AI tool leaving deidentified data vulnerable to cybertheft.

Below is a risk matrix showing risk of harm compared to degree of harm and recommending a level of disclosure for each scenario. Even the more lax disclosures should note the purpose, the very basic inner working, and the risks. No level of

[12] See Pilkington and Binkley; Rose and Shapiro.

disclosure for informed consent purposes should feel merely like a disclaimer. Level 1 should include the most detail about the AI tool itself. Disclosure level 2 is significant, but perhaps less detailed. Disclosure level 3 is the lowest level, and, in some cases, it could be a standardized line or two in a consent form. There are many samples of informed consent documents for AI in medical care.[13]

In evaluating the risk of harm, informed consent presumes that the AI tool itself is not so dangerous that its use is not medically appropriate. The high risk of devastation in the table below would be reserved for very special circumstances comparable to a life-threatening imminent situation and a daring surgical rescue. Some tools with a high likelihood of devastation would not have been purchased, FDA approved (if relevant), or in clinical use. However, as with risky cancer treatments like high-dose chemotherapy and stem cell transplants, clinicians offer, and healthcare consumers often accept, high-risk endeavors. For example, the risk of a blood clot during surgery or the risk of internal bleeding after surgery is often noted in consent documents. The relatively low risk of devastating consequences—even death—does not generally prevent consumers from undergoing surgery, especially for saving life or organs. In other cases, the riskiest aspect may be that the intervention does not work. A high likelihood of a low-impact event may be like the likelihood of nausea during chemotherapy. The disclosure would not be likely to provide significant detail about how and why chemotherapy causes nausea. Similarly, the disclosure about a common and mild issue associated with AI would not need to cover much detail about its inner working.

Risk and degree of harm

Event	Low Harm	Moderate Harm	Severe Harm
Highly likely	D3	D1	D1
Probable	D3	D1	D1
Low likelihood	D3	D2	D1

7.3.2 Step 2: Reliance compared to design

Where there is a risk of moderate to severe harm, the framework calls for analyzing overreliance. Clinicians may go along with the recommendations that AI systems supply due to automation bias. In cases where the AI eclipses the clinician experience or know-how, there is a strong case for detailed disclosure. Rather than looking just at the degree of autonomy in AI tools, the degree of clinician reliance on any tool is crucial. Often, decision support systems become the decision maker—clinicians

[13] CarePatron website. https://www.carepatron.com/templates/carepatron-ai-consent-form (Like numerous other tech providers, CarePatron has its own recommended informed consent document.); Bloomberg forms https://www.bloomberglaw.com/external/document/XF6H2E20000000/health-care-transactions-sample-document-patient-opt-in-consent-_ (generative AI consent form);

rely so heavily on decision support tools that the tool's decision then replaces the clinician's independent conclusion.

There are degrees of reliance and degrees of capability and appropriate use of any AI tool. In the matrix below, overreliance is not applicable to tools that are meant to be relied upon completely, i.e., autonomous tools. Such tools call for significant disclosures despite that proper use includes heavy reliance.

Influence on clinician recommendations

Intended reliance→ Risk of overreliance↓	Little reliance	Medium reliance	Heavy, definitive reliance
No risk of overreliance	D3	D2	D1
Low risk of overreliance	D3	D2	D1
High risk of overreliance	D1	D1	D1
Complete reliance is appropriate	D1	D1	D1

7.3.3 Disclosure Recommendations

Disclosure level 3 (D3) should include items 1-3; level 2 (D2) should include 1-4. And level 1 (D1) should include 1-7:

1. benefits and purpose (why use the AI tool?),
2. risks and limitations, including physical harm and data-related risks,
3. how data will be collected, used, and stored (if relevant),
4. error, accuracy, and the basis for any accuracy claims,
5. description of the AI design (training data, machine learning, supervised or unsupervised, generative, predictive, etc.),
6. degree of explainability, what is unknown about how the AI comes to its predictions, and
7. existence of protected trade secret/intellectual property

Disclosures should use plain language. Screen-based informed consent should include options to click for more information about an AI tool. Disclosures should put key information in a box, and use bullet points, bold type, and other visual aids. AI should not be buried in fine print.

7.4 Using the Framework

Case study. One quick fictional case study for using the framework is an oncology tool (let's call it CancerMatchAI) that analyzes tumor tissue for somatic alterations and identifies drug targets. Then the AI lists the drugs matching the targets. Clinicians who receive the final printout tend to show a bias toward the AI tool, confirmed by the number of times they prescribe a drug based on the AI. The tool is highly accurate and unlikely to make an error concerning the target expression or prevalence in the tumor. A healthcare consumer is deciding between an older chemotherapy agent that has harsh side effects and a newer precision medicine recommended by the AI tool.

Analysis. Beyond the scope of the AI, there is the added risk that even a drug precisely designed for the target may work for only a short time, or not at all, and the cancer may become resistant. The physical risks outside the AI include forgoing other interventions, going with the drug, and having it fail. The AI autonomously performs the analysis, but the clinician is meant to evaluate all other aspects of recommending a treatment plan from side effects to data about long-term survivorship. Clinical trials, institutional databases, and experience all contribute to the body of knowledge clinicians need for decision making. The AI tool is limited in that it makes the match, but the match itself does not mean that the drug will work for the healthcare consumer or that it is the best drug for the situation. The low risk of error is somewhat outweighed by the severity of the harm an error would cause. Even though the tool was designed for complete reliance as far as the tumor sequencing analysis, it was not designed to be the sole source of information for making a treatment plan. Even an excellent tool that autonomously performs complex bioinformatics calls for detailed disclosure. The analysis may lead to the conclusion 'italics' below.

Event	Low Harm	Moderate Harm	Severe Harm
Highly likely	D3	D1	D1
Probable	D3	D1	D1
Low likelihood	D3	D2	*D1*
Intended reliance→ Risk of overreliance↓	Little reliance	Medium reliance	Heavy, definitive reliance
No risk of overreliance	D3	D2	D1
Low risk of overreliance	D3	D2	D1
High risk of overreliance	D1	*D1*	D1
Complete reliance is appropriate	D1	D1	D1(*autonomous as to the task of identifying somatic alterations that are drug targets.)

Chapter 8
AI in Healthcare Settings

8.1 AI by Healthcare Setting

This section looks at AI in healthcare settings and explores considerations for applying the framework. In some cases, there would be no identifiable consumer for purposes of informed consent. Obviously, AI used for tasks unrelated to healthcare consumers, for example, Microsoft CoPilot used by clinicians to find a restaurant for lunch meetings, do not fall within informed consent frameworks. Similarly, calendar organizers or AI that alphabetizes would not require disclosure to healthcare consumers. And within healthcare uses, some disclosures for low-risk AI tools can be simple and succinct. Rather than case studies, and strict application of the framework, this section addresses the complexities of AI in healthcare settings.

8.2 Drug Discovery—Transparency

AI in drug discovery analyzes data to find patterns that may be overlooked by scientists, allowing the discovery of new molecules and drugs. AI tools also match diseases with existing drugs. AI can screen compounds quickly by simulating the "real disease." For example, AI can look at molecules in chemical compounds and simulate their effect on biological targets.[1] AI expedites questions that scientists ask, such as "what will bind to this protein?" or "which genetic characteristic must be present in a tumor for this drug to work?".

Generative adversarial networks (GANs) are two-pronged systems (two neural networks) whereby generated data is compared to and distinguished from real data. They use big data, and the two parts (the generator and the discriminator) work in competition with each other to improve output. The generator creates new data

[1] Abbas, M., Rassam, A., et al., (2024) The Role of AI in Drug Discovery. ChemBioChem (Wiley). doi.org/10.1002/cbic.202300816.

A. Zimmerman, *Informed Consent for AI and Health Care*,
SpringerBriefs in Ethics, https://doi.org/10.1007/978-3-032-11347-4_8

based on the training data—then, the discriminator classifies data as real or not. (The discriminator calculates the probability that data is part of the training data set.) In drug discovery, GANs are used to create chemical and biological datasets.[2] GANs help scientists find chemical compounds with desirable traits. Without GANs, scientists would not be able to identify each molecule as there would be a nearly infinite number of possible molecules in a "chemical space."[3] GANs allow testing variations of drugs for efficacy and side effects.

At the time of drug discovery, there is no identifiable healthcare consumer for informed consent purposes. Therefore, informed consent is not relevant until the time of clinical trials using human research subjects and then clinical care, when consumers of drugs are at the forefront personally. AI is not immaterial or unimportant; the absence of informed consent is due merely to the lack of an identifiable consumer. Disclosures about the drug development stage may be relevant in the clinical trial phase.

Nonetheless, companies developing drugs should disclose use of AI to the public and later, to the consumer, for several reasons. Disclosures need not include aspects of AI tools that may be protected intellectual property. First, pharmaceutical companies and researchers should be transparent about the possibility of explainability issues. (Why does drug X work for disease A?) Second, the public is interested when drug discovery financing comes from government seed money. People likely favor the use of AI in drug discovery. It presents an economical way to avoid clinical trials for drugs unlikely to prove satisfactory in general or in certain populations. This may save many animals and limit experimentation on humans. However, AI is not innocuous. When public institutions (the NIH) finance AI that leads to clinical trials for drugs or AI-based interventions that will help very few people or that are extremely expensive, the public should have some ability to weigh in. Voting and advocacy provide the public with some input. Companies often disclose uses of AI because innovation and promising solutions to diseases are a source of pride.

When something is faulty in the drug development AI, or when the data is biased so the resulting drug or intervention is not as effective for the general population, clinical trials may reveal the deficit. However, initial research reveals that drugs developed using AI (AI-discovered molecules) are more likely to succeed in clinical trials.[4] This is an area where there is significant potential.

Those whose data was used in creating the training sets and the AI models have different reasons for informed consent. As noted, unjust enrichment, risk of reidentification, and unanticipated use of data call for consumer protection.

---

[2] Tripathi, S., Augustin, A. I., Dunlop, A., et al., (2022). Recent advances and application of generative adversarial networks in drug discovery, development, and targeting. *Artificial Intelligence in the life Sciences*. https://doi.org/10.1016/j.ailsci.2022.100045 (includes overview of the mathematics and the architecture.)

[3] Tripathi, et al., 2022.

[4] Jayatunga, M. K., Ayers, M., Bruens, L., Jayanth, D., & Meier, C. (2024). How successful are AI-discovered drugs in clinical trials? A first analysis and emerging lessons. *Drug Discovery Today*, 104,009. https://doi.org/10.1016/j.drudis.2024.104009.

8.3 Clinical Trials

AI has many uses including study design, recruitment, screening candidates, and improving efficiency as well as diagnosis and matching research subjects to treatment regimens. Research subjects in early clinical trials have a case for a right to information about how AI was used in drug development. Research subjects generally sign off on physical risks like side effects rather than information about the drug development process. Especially because early clinical trials are experimental, the need to protect oneself requires significant disclosures. AI should be disclosed so that research subjects who oppose the use of unexplainable AI or who want more detail as to the type of AI and what it was used for can ask questions. At the point of clinical trial or even of recruitment, some disclosure about AI used in drug discovery is appropriate. Risks of clinical AI include failure to perform well in the clinic due to the strong fit to the training data, algorithmic bias, and privacy breaches.

8.3.1 Recruitment and Risk of Physical Harm

This is an area where healthcare consumers should have more information. A healthcare consumer may wonder, "Why me?" Individuals excluded from and those included in clinical trials have noted bias and outright discrimination. AI has potential to eliminate discrimination by focusing enrollment on medical criteria as well as potential to exacerbate discrimination by using past data reflecting those with access to more medical interventions. AI can wrongly eliminate people from clinical trials when algorithms rely on non-medical data, like insurance coverage for rehabilitation, or old data including who performed well w ith access to complex or cutting-edge care requiring wealth.

AI tools use complex algorithms to assess whether candidates meet the enrollment criteria for clinical trials. This can be very technical rather than a mere glance at diagnosis or criteria like disease progression and age, etc. For example, researchers explored the use of an AI tool to create tissue overlays, allowing researchers to categorize potential research subjects' liver disease.[5] Using human evaluation of research subjects, many research subjects initially thought to meet enrollment criteria did not. Introducing AI allowed researchers to better measure treatment response and failure and identify an appropriate population. In the study of the AI tool for liver disease, the AI assessed efficacy effectively as well.[6] While it covers a specific

[5] Iyer, J.S., Juyal, D., Le, Q. et al. (2024). AI-based automation of enrollment criteria and endpoint assessment in clinical trials in liver diseases. *Nature Medicine* 30, 2914–2923. https://doi.org/10.1038/s41591-024-03172-7 (Tool to assist "pathologists in histologic review of MASH clinical trials." The FDA is evaluating MASH as a drug discovery/clinical trial tool.)

[6] Iyer, et al., 2024.

disease, the AI tool led to reproducible, accurate assessments, and the tool looks promising for use in clinical trial recruitment.[7]

AI tools for recruitment assess the potential that things will go wrong. They identify risky research subjects and they fine tune trial participants to those more likely to thrive.[8] One tool, Trial Pathfinder, was able to double the number of eligible clinical trial participants for a lung cancer study without increasing the rates of negative incidents. The AI tool can open clinical trials up to many who would have been excluded without the tool, demonstrating the tool was better at assessing which participants could withstand negative side effects.[9]

An AI tool for Alzheimer's disease research has identified participants using predictive models. AI tools have identified amyloidosis potential based on MRI and family history.[10] AI using biomarkers has the potential to decrease clinical trial failure. Researchers suggest, "This incredible use of AI can help reduce recruitment time for patient selection, reduce unnecessary screening procedures, predict the most suitable trial patients, and improve trial effectiveness."[11]

Because potential research subjects are often healthcare consumers—they are sick and need intervention—exclusion from a clinical trial can mean life or death. While bioethicists often discuss "therapeutic misconception", some sick healthcare consumers with end-stage or devastating cancers, for example, undergo an unproven intervention partly or entirely because it might work for them. Research subjects should (and do) understand that their experience and results will contribute to researchers' scientific knowledge and development of treatments for others.

Researchers should obtain informed consent noting the use of clinical trial recruitment AI. Research subjects arguably want to know why they were recruited. Because medical research has been both predatory[12] and exclusive,[13] there are many ethical reasons to require disclosures. To generalize, the harm caused by being wrongly excluded from a clinical trial is quite high, especially if in hindsight the intervention proved efficacious. Those trying to get in should be able to obtain information about

[7] Iyer, et al., 2024 (AIM-MASH is a group of AI algorithms to "predict the location, extent and severity of histologic biomarkers of MASH" (metabolic dysfunction-associated steatohepatitis), a severe form of what was known as nonalcoholic fatty liver disease.)

[8] Hutson, M. (2024) Hutson, M. (2024). How AI is being used to accelerate clinical trials: From study design to patient recruitment, researchers are investigating ways that technology could speed up the process. *Nature*, 627(8003), S2-S5.

[9] Hutson, 2024.

[10] Theories on the cause of Alzheimer's disease vary – this mention just notes the role of AI.

[11] Angelucci, F., Ai, A., Piendel, L., Cerman, J., & Hort, J. (2024). Integrating AI in fighting advancing Alzheimer: Diagnosis, prevention, treatment, monitoring, mechanisms, and clinical trials. Current Opinion in Structural Biology, 87, 102,857. https://doi.org/10.1016/j.sbi.2024.102857.

[12] Clinton, W. (1997) Remarks by the President in Apology for Study Done in Tuskegee. https://clintonwhitehouse4.archives.gov/textonly/New/Remarks/Fri/19970516-898.html (President Bill Clinton formally apologized for ethics violations and racism in the Tuskegee syphilis study.)

[13] Hoffman, S. (2000). Beneficial and Unusual Punishment: An Argument in Support of Prisoner Participation.

In Clinical Trials. *Indiana Law Review*. Vol. 33: 475. https://mckinneylaw.iu.edu/practice/law-reviews/ilr/pdf/vol33p475.pdf (Example of prisoner exclusions.)

eligibility. If AI is used, full disclosure about its use would be appropriate. Because those excluded do not sign the consent documents that those included do, they should be entitled to an explanation.

An AI tool including someone inappropriate for a clinical trial could also cause grave harm. For example, assigning chemotherapy to a person unlikely to benefit or likely to succumb to side effects would be a recruitment error. Informed consent acts as one additional check on recruitment of individuals who are not a fit. Knowing AI was used to recruit research subjects is material to some subjects who want to know which factors including their symptoms, genetics, disease stage, or other characteristics led to their inclusion.

Clearly some clinical trials, for example, one assessing the benefits of an electrolyte drink prior to surgery, do not carry high risk and a smaller disclosure about any AI involved would be adequate. The inclusion in the trial and the leaving out pose little risk. The disclosure of the use of AI could then be level 3 as without a high risk, a simple standard line about the use of AI should suffice. If the liver assessment tool described above were implemented, someone incorrectly included in a clinical trial could pass up a more appropriate disease management program and be harmed significantly. The disclosure should be level 1 (the most detailed), or level 2 based on the potential harm.

The second matrix, disclosure as it relates to degree of reliance on the AI tool, also supports moderate to highly detailed disclosures for recruitment AI. Arguably, researchers are highly likely to go along with AI in enrollment. Some clinical trials fail due to having too few enrollees. And clinicians who see trials as a way to help their healthcare consumers may be biased toward an AI recruitment tool that suggests the healthcare consumer is a good fit. So far, some tools have proven more reliable than non-AI evaluation of enrollment eligibility. The AI tools are not meant to be relied on completely—they are essentially decision support tools in many cases. After applying the complex AI tools, researchers could still use other criteria like location, severity of condition, or even ability to pay. The risk of automation bias is high. *A good ethical rule is that if the researcher is likely to go along with the AI tool without significant individual consideration of the research subject, then the amount of reliance (overreliance) may exceed the intended reliance.* Such a situation should call for a more comprehensive disclosure. Taking the liver example, the AI tools are complex, and the likelihood of reliance is high. If the tool is proven as helpful as anticipated, researchers would not be able to assess the probability of success for each research subject as well, efficiently, or quickly without the AI. If the risk of overreliance is high and the tool is designed for medium reliance with appropriate oversight or weeding out based on some criteria that is not embedded in the AI, then more detailed disclosure would be appropriate. FDA approval of the tool should indicate it is highly accurate and consistent. (FDA approval of software as a medical device would indicate the tool is not for just clinical decision support.) However, generally if a tool recommends a participant to a clinical trial, the tool itself does not accept or reject the research subject. AI tools merely give the medical indication of whether the candidate is a good fit. If a research subject were to ask, "Why am I a good fit for this clinical trial?" the researcher should describe the role of AI and how the AI impacted the researcher's decision to include the participant.

Generative AI to improve healthcare consumer education about clinical trials can make it easier for research subjects to understand clinical trials.[14] LLMs summarize complex documents and prepare Q&A sections to ensure research subjects understand the intervention, risks, and benefits. However, the risk of hallucination or any incorrect information in a summary is high enough to require human oversight. Generative AI has the potential to improve informed consent by presenting understandable, jargon-free explanations. Yet humans should read over and edit the final documents.

8.3.2 Interventions and Risk of Physical Harm

Once recruited, research subjects could encounter AI-based recommendations and predictions throughout the clinical trial. AI may transform clinical trials from the status quo of one-size-fits-all to a personalized precision medicine approach. By constantly analyzing drug performance and side effects,[15] AI may contribute to dynamic clinical trials. This upsets the typical protocol style pursuant to which individuals are treated the same way so that efficacy can be assessed across a healthcare consumer population with the same disease. Precision medicine allows personalized treatment plans. Clinical trials have been reimagined to accommodate precision medicine. By clumping people who are the most alike by biomarkers yet have different cancers (basket trials) and by allowing personalized treatments for the same disease (umbrella trials), researchers have found ways to expand precision medicine beyond "n = 1" trials. GANs, described above, are used in many types of clinical trials. For example, Alzheimer's classification relying on AI generated synthetic PET images resulted in higher accuracy than classification without the use of synthetic PET images.[16] Similarly, trials may take a person's tumor tissue and create an individualized "vaccine" approach. The substance delivered to each healthcare consumer is not uniform across participants although the methods are.

The type and degree of risk is high in many clinical trials. In addition to expected treatment risks, the experimental nature increases the possibility of unknown risk. Disclosure about the use of AI may be material to decision making. Evaluating the harm and likelihood of the harm (the first matrix) as well as the degree of reliance on AI and the intended degree of reliance on it (the second matrix) would help researchers determine how detailed a disclosure must be to meet the requirements of informed consent.

---

[14] Waters, M. (2025) AI meets informed consent: a new era for clinical trial communication. JNCI Cancer Spectrum, 2025, 9(2) https://doi.org/10.1093/jncics/pkaf028.

[15] Angelucci, et al., 2024.

[16] Angelucci, et al., 2024.

8.4 Software as a Medical Device – Radiology and Risk of Physical Harm and Data-Related Harm

AI tools in radiology have been around performing tasks like sorting and categorizing. Radiology AI tools are trained on images and data. AI has reduced signal to noise ratios and improved radiologists' efficiency.[17] The AI may catch something the radiologist would not. All software that processes an image or signal will be considered software as a medical device and require FDA pre-market clearance. Changes at the FDA could change the regulatory landscape, making approval less predictable, and potentially changing the conditions under which the FDA requires AI tools to be submitted for approval.

When radiologists use AI to review MRI images, the consent to participate in the MRI does not consistently reveal the use of AI in reviewing scans. A radiologist is generally responsible for reading and signing off on radiology reports. A radiology report should disclose the use of AI for the sake of transparency. When using AI is planned, built-in to the current system, and both the radiologist and the other clinicians involved know about it, it would be easy to add its use to a consent form. There is already a standard consent form normally focused on risk of radiation for x-rays, risk of magnetics interacting with metals for MRIs, and risks posed by contrast.

There are regulations protecting healthcare consumers from AI or offshore companies producing radiology reports which radiologists merely rubber stamp. The radiologist would be liable for relying on faulty, inaccurate AI, and, in some cases, for failure to contextualize, etc. Yet AI systems nonetheless are likely to be very accurate and informative.

Radiology AI carries risks including privacy breaches, reidentification (with its associated risks of extortion, impact on ability to get insurance, employment, and subjecting the data subject to targeted advertising), unjust enrichment, error due to bias in the training data (over-fitting or selection bias), and variations when training data combines data from different machines (variation in MRI field strength or the slice the CT machine is set to image).[18] Furthermore, the ground truth for labeled datasets may be prone to error based on the decisive way AI sorts compared to the murky world of diagnostics. Arguably, many diseases cannot be diagnosed by one picture alone.[19]

Some suggest that a simple disclosure on the record describing the use of AI may cover it. For example, reports could add that "AI was used in comparing your radiology images to those in a large dataset." Yet, in many cases it would be appropriate to disclose significantly more about the AI tool. With a serious disease like

---

[17] Willemink, M.J., Noël, P.B. The evolution of image reconstruction for CT—from filtered back projection to artificial intelligence. *Eur Radiol* **29**, 2185–2195 (2019). https://doi.org/10.1007/s00330-018-5810-7.

[18] Brady, A. P., & Neri, E. (2020). Artificial Intelligence in Radiology-Ethical Considerations. *Diagnostics (Basel, Switzerland), 10*(4), 231. https://doi.org/10.3390/diagnostics10040231.

[19] Brady and Neri, 2020.

cancer, where scans become nearly routine, the harm from misinterpretation could be more severe. For example, a cancer patient may benefit from an AI tool relating images to disease progression, but overall care, diagnosis, and prognosis depend on information beyond the scan. The limitations and inner workings of an AI tool are highly relevant. For those not yet diagnosed at all, something being missed or misdiagnosed would be low probability but high impact. The harm matrix may point to greater, detailed disclosure; a reliance matrix is tricky. Some AI tools used as medical devices are designed for reliability. Yet clinicians will be the responsible parties and should engage in oversight. Automation bias is also highly relevant to AI in radiology. Analyzing any situation under both matrices, the more serious the underlying condition or the symptoms leading to the scan, the stronger the duty to disclose physical risks.

8.5 Digital Twins

Digital twins use machine learning and GANs.[20] They are used in drug discovery several ways. They can lead virtual clinical trials in which researchers learn about drugs without human and animal experimentation.[21] The model might focus on biological elements, physiology, or proteins, for example.[22] Digital twins allow researchers to analyze what disease progression would look like and to test interventions. Digital twin technology can replicate cells, systems, patients, and populations using numerical representations and computations.[23]

At the clinical level, digital twins are personalized replicas representing the patient. They can combine many types of data using multimodal models. The patient's data is kept current as new data arises through patient monitoring.[24] Digital twin products are designed to more accurately pinpoint appropriate treatments

---

[20] Li, T., Long, Q., Chai, H., et al., (2025). Generative AI Empowered Network Digital Twins: Architecture, Technologies, and Applications. ACM Computing Surveys, 57(6), 1–43. https://dl.acm.org/doi/full/10.1145/3711682.

[21] Wang, H., Arulraj, T. Ippolito, A., and Popel, A. (2024). From virtual patients to digital twins in immuno-oncology: lessons learned from mechanistic quantitative systems pharmacology modeling. *npj Digital Medicine*. https://www.nature.com/articles/s41746-024-01188-4.

[22] Laubenbacher, R., Mehrad, B., Shmulevich, I. et al. (2024). Digital twins in medicine. *Nature Computational Science* 4, 184–191. https://doi.org/10.1038/s43588-024-00607-6.

[23] European Organisation for Research and Treatment of Cancer (2024). Scientists create cancer patients' 'digital twins' to predict how well treatments may work. *eCancer.* https://ecancer.org/en/news/25568-scientists-create-cancer-patients-digital-twins-to-predict-how-well-treatments-may-work.

[24] Mollica, L., Leli, C., et al. (2025). Digital twins: a new paradigm in oncology in the era of big data. *Real World Data and Digital Oncology*. https://doi.org/10.1016/j.esmorw.2024.100056 (describing digital twin technology using big data and personalized data.)

for complex cases.[25] In oncology, digital twins allow clinicians and researchers to simulate the progression of cancer in the healthcare consumer. Cardiac digital twins allow improved modeling of heart disease and can lead to better, personalized management.[26] Digital twins and their granular prediction about treatment are promising.

Digital twins are not without risk. Biased data or a lack of diversity in the data could lead to a recommended treatment that is not beneficial. When using a digital twin, a healthcare consumer could forgo a standard treatment option in the hope that the highly precise outcome would be beneficial. Continuously updating the model is important. There are some challenges to creating and using digital twins including the following:

- Collecting high-quality and high-volume data derived from diverse populations,
- Optimizing mathematical and AI-based models according to relevant outcomes
- Reducing risk of learning from biased data
- Improving the continuous update of data from real-world resources
- Safeguarding data privacy and security
- Ensuring equitable representation of diversity in terms of gender, ethnicity, and social background[27]

The physical risk includes the opportunity cost of forgoing an established treatment that is often the standard of care. The rigorous testing of digital twin products and proof of positive outcomes is crucial to their use. The scientific community including the FDA, NIH, and National Science Foundation viewed digital twins as promising technology worthy of funding,[28] although it is unclear whether digital twin research will survive federal funding cuts.

Digital twin products perform their task of recommending optimal treatments nearly autonomously. If the digital twin technology made an oddball recommendation, researchers and clinicians could intervene and see whether their knowledge and expertise support the recommendation. Clinicians should understand the treatment options and the research subject's medical history and comorbidities, and have additional information or context to prevent automation bias from leading to just going along with the output when following a human advice in the form of a clinician-devised treatment plan may be beneficial. There is a strong risk that a clinician or

[25] FarrSight-Twin technology transforms cancer treatment research. *News-Medical.net*. https://www.news-medical.net/news/20241024/FarrSight-Twin-technology-transforms-cancer-treatment-research.aspx ("Each digital twin is created from biological data from thousands of patients with cancer who have been treated in different ways. This information is combined to recreate the cancer of a real patient with molecular data on their tumor. This digital twin makes it possible to predict how a patient is likely to respond to a treatment.").

[26] Laubenbacher, R., Mehrad, B., Shmulevich, I. et al. (2024). Digital twins in medicine. *Nature Computational Science* 4, 184–191. https://doi.org/10.1038/s43588-024-00607-6.

[27] Mollica, et al., 2025. (quotation.)

[28] NSF, NIH and FDA support research in digital twin technology for biomedical applications. US National Science Foundation. Website. https://www.nsf.gov/news/nsf-nih-fda-support-research-digital-twin-technology.

researcher would rely nearly exclusively on digital twin technology. Digital twins are so personalized that researchers may be overconfident that the person would behave or react to interventions the same way as the digital twin. As far as automation bias, digital twin technology is a breakthrough—clinicians may be impressed with what it offers and may be unsure of how to weigh their experience, or in some cases a less-than-satisfactory standard of care, against the digital twin's recommendation. This is an area where disclosure is important to ensure that innovation is not stifled by ethics-related failures like failing to provide the healthcare consumer with enough information.

8.6 Decision Support Tools and Risk of Physical Harm

Clinical decision support tools are distinguished from software as a medical device. The FDA carves out exceptions to the rule that software as a medical device, for example all imaging AI and AI that provides determinative output rather than listing recommendations, must have approval. The clinical-decision-support exception covers some AI that will not process images. Many clinical reminders and recommendation systems, as long as they are not too specific (rather than being directive, they should give options), do not require FDA approval. Yet escaping stricter regulatory scrutiny does not imply that consumers do not need to understand them when they agree to a treatment. They are often part of the EHR system.

As in the Practice Fusion case, there can be wrongdoing in the design of the AI tool. And just by typical training data, AI could promote drug use due to training data from a medicalized online community. Clinical decision support tools are in some ways stealth—the FDA is less concerned about their use, and they are built into the practice of medicine. However, disclosures about clinical decision support tools are crucial. Consumers do not realize the influence of reminders, pop-up alerts, recommendation systems, and tools providing guidance to clinicians. They operate on the screen and are not consumer facing.

Decision support tools can be harmful. They can recommend drugs with serious side effects, be biased, have explainability issues, and contribute meaningfully to dosing, prescribing, diagnostic, and other errors. However, they can reduce errors, improve efficiency, and ease the delivery of care. They are used in triage, emergency care, and other settings in which quick decisions are crucial. Some AI tools vary in their accuracy based on how and when they are used—one tool for sepsis was accurate in identifying people likely to develop sepsis only when initial tests were already performed. The tool was much less accurate when no suspicion of sepsis had yet arisen.[29] The market for AI triage tools is growing and emergency settings may benefit significantly from the use of AI.

---

[29] Smith, D. (2024) Widely used AI tool for early sepsis detection may be cribbing doctors' suspicions. Michigan Engineering. https://news.engin.umich.edu/2024/02/widely-used-ai-tool-for-early-sepsis-detection-may-be-cribbing-doctors-suspicions/

Decision support tools also make recommendations about discharge decisions. When such tools are biased, they may make recommendations that are inappropriately informed by demographics, including race, income, and livelihood rather than on actual data about whether one's apartment has stairs or whether the healthcare consumer has a caregiver. Demographic data in the training set is used as a proxy and AI harnesses correlations, making assumptions about consumers based on others who are like them in some arguably not meaningful ways.

Clinicians have a high likelihood of overreliance on them.[30] They are specifically designed to assist rather than make decisions. Clinicians now use generative AI tools for decision support as well. The disconnect between intended reliance and actual reliance is likely to be especially high as decision support tools are specifically intended to stop at assistance. They are marketed specifically as just support. They escape regulatory scrutiny because they are defined as not intended to make medical decisions. While it depends on the situation, AI tool, and the use, both matrices point to relatively significant disclosure.

8.6.1 Administrative AI and Risk of Data-Related Harm

Administrative AI is becoming ubiquitous in health care. Large systems share information among and within entities and AI tools are involved in a number of ways. The importance of EHR systems has changed significantly. Now that the systems include recommendations, alerts and reminders, and generative AI tools, EHR systems can impact healthcare consumer outcomes and experiences more significantly. (The clinical tools built into the AI systems are discussed above under clinical decision support.) Schedules, health status, prescription drugs, demographics, billing, appointment notes, and insurance information are housed together on systems. While encryption and other precautions are often required, the systems pose risks, even beyond privacy and cybersecurity. The disclosures to healthcare consumers should cover both AI that eases administrative tasks and other AI tools that may be trained on the consumers' data.

When paperwork was handwritten and did not follow healthcare consumers to new hospital systems or medical practices, the impact of an error was different. Now single errors may be seen by and transferred among institutions. Shared systems allow clinicians to correct errors across the platform, yet erroneous data may already have been incorporated into large datasets. As noted, there are privacy, cybersecurity, and other risks. Identifiable protected health information is abundant in electronic records. A leak of a stigmatized health condition or a failure to pay for care can lead to reputational damage or extortion. A cybersecurity breach can lead to financial

[30] Saadeh, M.I., Janhonen, J., Beer, E., Castelyn, C., Hoffman, D. (2025). Automation complacency: risks of abdicating medical decision making. *AI Ethics*. https://doi.org/10.1007/s43681-025-00825-2.

damage. With social security numbers and other personal information, cybercriminals can access bank accounts. Some suggest that the ethical use of data depends on telling the consumer the financial value of the data.[31] However, the consumer signing off may be deemed to have waived any rights to compensation.

Whether a reasonable person would want to know about AI used in the system may depend on what AI is used for and whether a healthcare consumer would change behavior, provide less data, or seek care elsewhere. Then applying the framework, harm caused by inappropriate accessing of consumer data is not physical but may lead to serious consequences. The potential for harm compared to the degree of harm depends partly on the system itself—how secure it is, which employees can access it, and how information in it is passed along to outside entities. It also depends on accurate data entry. The risk of leaking data that identifies the consumer calls for reasonable disclosure. When the risk is privacy or cybersecurity, the level of detail about how the AI tool works or comes to decisions may be less important than assurances that information will be kept safe. There are a regulatory framework and strict requirements for healthcare cybersecurity.

The reliance on AI may be appropriate to its design. It is likely that clinicians take things entered into EHRs seriously—they view them as factual and likely are not concerned about errors or that many items in records are opinions or judgments rather than facts. The EHR and other administrative AI tools are designed for relatively high reliance. Institutions purchase the systems and then clinicians are not expected to add to protecting privacy or cybersecurity of the data after it is entered. Applying the reliance matrix requires asking whether a clinician would view the data in the EHR as an accurate representation of the healthcare consumer's condition, status, demographics, etc. The risk of human error in data entry is a concern. Rights to correction and deletion exist regardless of consent.

Data from EHR systems can be used to train AI tools. Before any data is entered about a consumer, the healthcare consumer should be informed about how the data could be used in training AI, marketing, and medical research, etc. With broad use clauses slipped in, consumers may not know how their data contributes to AI tools or medical endeavors. Some consumers may not want their data used to train tools that facilitate research that they find morally offensive or wasteful. Others may be especially concerned about unjust enrichment. Research is mixed as to consumer willingness to share health data with third parties.[32] These harms are difficult to rank or value. Many would consent to the use of data for medical research that is likely to benefit society or those with serious illnesses. If the tool were used appropriately, cybersecurity were airtight, and the data would not be used for any unspecified purposes or sold to data brokers, a less detailed disclosure (level 3) may be enough. For the data-related harms, including a breach of privacy, the institutional

[31] Brady, A. P., & Neri, E. (2020). Artificial Intelligence in Radiology-Ethical Considerations. Diagnostics (Basel, Switzerland), 10(4), 231. https://doi.org/10.3390/diagnostics10040231.

[32] Trinidad, M., Platt, J. & Kardia, S. (2020). The public's comfort with sharing health data with third-party commercial companies. *Humanities and Social Sciences Communications* 7, 149. https://doi.org/10.1057/s41599-020-00641-5.

level protections should protect healthcare consumers, but nonetheless they should have information about how their data is collected, stored, transferred, sold, and used. Informed consent is highly relevant to the data associated with the EHR.

8.6.2 Generative AI and Risk of Data-Related Harm

Incorrect records may result from generative AI. This may cause varying degrees of harm. AI tools designed to summarize notes can make errors like hallucination, emphasize that which the clinician would have deemphasized, and leave out important points. Some AI tools like Google Notebook LM have visible errors in their generated conversations based on documents. An AI-generated conversation about three or four documents often ascribes an incorrect viewpoint to an author or mischaracterizes a claim by adding confidence and emphasis.

In a Texas consumer protection case, Attorney General Ken Paxton reached a settlement with Pieces Technologies, a company that makes technology to generate summaries of healthcare consumer's health conditions and treatments for inpatient facilities.[33] The settlement agreement that was reached (a court-approved assurance of voluntary compliance), set forth five criteria that must be disclosed to purchasers. They are (1) the type of data used to train the model and the type of model, (2) its intended purpose and any instructional or training materials for users, (3) "any known, or reasonably knowable, limitations of its products or services, including risks to healthcare consumers and healthcare providers from the use of the product or service, such as the risk of physical or financial injury in connection with a product or service's inaccurate output", (4) and known or knowable limitations and risks of harm (physical and financial), and (5) a catchall: "for each product or service, all other documentation reasonably necessary for a user to understand the nature and purpose of an output generated by a product or service, monitor for patterns of inaccuracy, and reasonably avoid misuse of the product or service."[34] These disclosures would add transparency and protect purchasers thereby, somewhat indirectly, protecting healthcare consumers. However, the order did not touch on whether hospitals would or should disclose the same information to the healthcare consumers whose visits/medical histories, etc. they were summarizing using the AI. Accuracy and hallucination rates were the primary issues in the case. The attorney general claimed the AI tools were less accurate than advertised.

---

33 Texas Attorney General's Office. (2024) Attorney General Ken Paxton Reaches Settlement in First-of-its-Kind Healthcare Generative AI Investigation (press release). https://www.texasattorneygeneral.gov/news/releases/attorney-general-ken-paxton-reaches-settlement-first-its-kind-healthcare-generative-ai-investigation.

34 Texas v. Pieces Technologies, Inc. Petition for Approval and Entry Of.
Assurance Of Voluntary Compliance. 2024. https://www.texasattorneygeneral.gov/sites/default/files/images/press/Petition%20for%20Approval%20of%20AVC%20Pieces%20File%20Stamped.pdf.

The Texas case demonstrates that informed consent is nearly ignored, or "off the table." The court-ordered transparency does not extend to the doctor-patient relationship. It lives in the purchaser-seller relationship, where it provides an important precedent for consumer protection. The Texas Attorney General was acting on behalf of purchasers and the public. This is the language of the principle of transparency in the legal setting of consumer protection. This case protects the consumer of hospital care, but only by protecting the purchaser of the product.

The likelihood of harm is based on rigorous testing of the AI tool's accuracy, hallucination rate, and known errors. As with all AI tools, clinicians should not be using tools that are error-prone and unreliable. Generative AI is already in use in clinical-care information gathering. It does not go through a regulatory preapproval process. Clinicians may be underestimating the potential harm and overvaluing the degree to which AI relieves the burden of entering notes. In addition to the risks of privacy-related or physical harm, generative AI could undermine trust in the doctor-patient relationship.

Generated notes carry physical risks as well—they can lead to incorrect prescribing or misdiagnosis. Without AI, clinicians and administrative staff can enter erroneous notes. If an AI tool increases error, it arguably should not be used. For appropriate AI tools, consumers should know about use, purpose, and risks. Because generative AI is new, it is worth disclosing some details about how it works to those consumers who may not understand natural language processing and large language models. Assessing the degree of harm compared to the likelihood would include some analysis of oversight, physical, reputational, and privacy risks.

Generative AI tools are often linked to overreliance. In a lawsuit, AI generated citations to nonexistent cases. Generative AI hallucinations have ranged from comical to dangerous.[35] Partly because they tend to provide a shortcut to creative, research, and writing tasks, there is a strong tendency to overly rely on them and to skip the human oversight necessary to ensure they are correct. In many fields, humans take advantage of AI—this can be seen as cheating, especially when generative AI writes a paper, or as a helpful tool when it outlines or summarizes. The high risk of overreliance while the potential for mistakes is quite high calls for more disclosure when evaluated using the reliance matrix.

8.7 Ambient AI

Ambient AI operates in the background, usually continuously taking in data from an environment or person. For example, "smart homes" that use AI to adjust lighting, temperature, and sound are ambient systems. In healthcare settings, ambient AI refers to technology that combines voice recognition, chat and virtual assistant technology,

---

[35] Gautam, A. (2024) Oops, My AI Did It Again: The Funny (and Bizarre) World of AI Hallucinations. LinkedIn. https://www.linkedin.com/pulse/oops-my-ai-did-again-funny-bizarre-world-amrita-gautam-oa3ic/

facial expression analysis, and monitoring. AI nurses are an example of ambient AI. Current products do not entirely account for body language, sarcasm, sadness, and regret. Ambient AI can also streamline the use of EHR systems, working both directions, to add to a record and pull information from it.

In a German study, ambient technology was installed in an oncology waiting room to perform real time emotion assessments. The research subjects provided informed consent. They stood in front of a monitor which assessed them. The researchers noted that people were amenable to the collection of information and felt it may be useful.[36] The AI assessed emotional state by analyzing facial expressions. Ambient technology could be a first step in a triage process. In the study mentioned, the overt nature of the screen and the informed consent process contributed to its ethical foundation.

Ambient technology provides one of the most pressing cases for requiring informed consent. When they know how it is used, both physicians and healthcare consumers may be more judicious about what they say, wear, and do. Ambient technology can invade privacy, eavesdrop, create a narrative that is added to the EHR, diagnose emotional and mental disorders, predict cognitive decline, recommend drugs, and collect and record significant amounts and types of data.

IEEE guidelines suggest that wearable clinical Internet of Things (IoT) AI tools should conform with standards of Trust, Identity, Privacy, Protection, Safety, and Security.[37] Informed consent is warranted to meet the concept of trust. Healthcare consumers wearing devices at home must be aware of what type of information a clinical IoT tool can pick up.

Evaluating it under the recommended framework for assessing AI, ambient care may not seem high risk, but it is. First, in the case of collecting and monitoring vital signs, any error or even a loss of electricity powering it could pose an issue. Second, many people seeking medical care do not want cognitive or psych assessments. And data collected in a waiting room or in an exam room but unrelated to the physical issue at hand could be used to promote interventions. Third, when vital signs become data, there is always a risk. For example, ambient technology can predict the likelihood that someone is pregnant quite accurately. In places with restrictive laws or in situations where a parent or significant other has access to medical records, pregnancy data poses danger. People hold strong opinions on pregnancy out of wedlock and abortion, and some states have very aggressive fetal assault laws that pose danger to pregnant women. Women have many reasons to avoid ambient technology picking up a pregnancy.

While it is not the prevailing opinion, the public generally should not always be under the gaze of medicine. Scholars have noted the way in which physicians

---

36 Binz, E., Seuß, D., Pahl, J., Ko, Y. D., & Wittenberg, T. (2023, September). Machine-based emotion-assessment in waiting rooms–a feasibility and acceptance study. In Current Directions in Biomedical Engineering (Vol. 9, No. 1, pp. 113–117). De Gruyter.

37 IEEE.org, Standards Association. IEEE/UL Standard for Clinical Internet of Things (IoT) Data and Device Interoperability with TIPPSS–Trust, Identity, Privacy, Protection, Safety, and Security. https://standards.ieee.org/ieee/2933/7592/

create patient stories using a medical model.[38] The AI itself may reflect the medical gaze that many sociologists and experts see contribute to medicalization. Even by interpreting body language and assigning emotion, AI may extend the reach of the psych industries and result in prescribing drugs to people who did not seek out a mental health assessment.

The harms associated with ambient AI may be privacy related or emotional, and in some cases physical. Each ambient AI tool evaluated may require different levels of disclosure. For example, ambient tools "watching" ICU patients may have more physical risk than ambient tools assessing nervousness among people in a waiting room. Reliance on the technology varies as well. If an ambient tool is replacing a nurse,[39] the stakes are likely high. If it is assessing cognitive function and could be used to curtail rights or freedom, that too is high risk. High reliance on such a tool could be ethically problematic. However, if ambient AI is not impacting care and is used to design a better examination room or improve comfort, informed consent may be irrelevant.

8.8 Insurance Authorization and Risk of Physical Harm—Beyond Informed Consent

Rather than (or in addition to) informed consent, regulation is needed to ensure healthcare consumers are treated fairly. Forty states have pending laws that require a human to be involved in coverage denials. AI should not be used alone to deny preauthorization of a treatment or to deny coverage of prescription drugs. Informed consent to AI is not likely to provide enough protection to healthcare consumers in need of authorization. Signoff to acknowledge the use of AI may be interpreted to indicate willingness to accept the outcome.

United Healthcare has been sued due to its use of an AI tool called nH Predict to deny claims. Plaintiffs claim that use of the tool violated contractual obligations and obligations of good faith. The plaintiffs allege that United Healthcare instructed claims agents to deny claims and keep any approved care to within one percent of the algorithm's recommendation. (The agents relied heavily on the AI tool, as required by their employer.) Ninety percent of denials that were appealed were reversed, likely indicating the algorithm was too strict in denying care that consumers really needed. The plaintiffs are Medicare Advantage plan customers who need rehabilitation and

---

[38] Misselbrook D. Foucault. Br J Gen Pract. 2013 Jun;63(611):312. https://doi.org/10.3399/bjgp13X668249. PMID: 23,735,390; PMCID: PMC3662436.

[39] Pearl, R. (2024). *Forbes*. https://www.forbes.com/sites/robertpearl/2024/04/17/nvidias-ai-bot-outperforms-nurses-heres-what-it-means-for-you/

post-acute care.[40] CMS has since clarified that it prohibits the use of AI to deny coverage[41]:

> An algorithm or software tool can be used to assist MA plans in making coverage determinations, but it is the responsibility of the MA organization to ensure that the algorithm or artificial intelligence complies with all applicable rules for how coverage determinations by MA organizations are made. For example, compliance is required with all of the rules at § 422.101(c) for making a determination of medical necessity, including that the MA organization base the decision on the individual patient's circumstances, so an algorithm that determines coverage based on a larger data set instead of the individual patient's medical history, the physician's recommendations, or clinical notes would not be compliant with § 422.101(c).[42]

The case of insurance authorization and denial highlights the insufficiency of informed consent to resolve all ethical issues caused by AI. A signoff is not enough to ensure ethical behavior. Keeping healthcare costs down and insurance company profits up through claim denials is a practice that the public and the policyholders should challenge regardless of how AI is used in the task. Uncertainty, waste, differing opinions, and some over-referral to rehab facilities are possible even when humans make the decisions. However, a common complaint about insurance companies is their sometimes-ruthless denial of claims and their interference with the ability of healthcare consumers to follow their physicians' recommendations.

While informed consent would highlight to the public that insurance companies are using AI this way, it is unlikely that people would find it easy to shop around for plans not using AI. To give an informed refusal of the AI while accepting the rest of the contract is an unlikely outcome based on the size and power of insurance companies. Therefore, regulation may be the best route to protect the public from algorithms designed to deny claims inappropriately. Both requiring human oversight and regulating the use and design of technology would achieve appropriate protection.

The high impact of the harm caused by a refusal to cover necessary treatment and the known heavy reliance on AI in the insurance industry call for serious disclosures about how these tools are trained, how they work, and how corporations use them. Informed consent to a non-outcome-determinative algorithm when purchasing health insurance would be helpful and could become standard or required

---

40 Lokken v. United Health Group, Inc. United States District Court, District of Minnesota. Memorandum opinion and order granting in part and denying in Part defendants' motion to dismiss (2025). https://www.courthousenews.com/wp-content/uploads/2025/02/UHG-judge-dismisses-counts-opinion.pdf (allowing claims of breach of contract and breach of the implied covenant of good faith and fair dealing; dismissing unjust enrichment, deception, and unfair competition claims.)

41 Centers for Medicare and Medicaid Services (CMS) (2024) Frequently Asked Questions related to Coverage Criteria and Utilization Management Requirements in CMS Final Rule (CMS-4201-F) CMS Final Rule. Memorandum. https://cdn.arstechnica.net/wp-content/uploads/2024/02/cms-memo-2624-faqs-related-to-coverage-criteria-and-utilization-management-requirements-in-cms-final-rule-cms-4201-f.pdf; Mole, B. (2024) AI cannot be used to deny health care coverage, feds clarify to insurers. ARS. Technica. https://arstechnica.com/science/2024/02/ai-cannot-be-used-to-deny-health-care-coverage-feds-clarify-to-insurers/

42 CMS, 2024, memorandum.

by law. Transparency and media attention have highlighted this issue and should bring about stricter regulatory oversight. Lawsuits also play an important role in holding companies accountable for unfair practices.

Chapter 9
Conclusion

9.1 Limitations and use of framework

To be ethically deployed, AI must meet many standards. Informed consent is not the only ethical requirement. For example, AI must be right for the situation, unbiased or its biases noted and accounted for, and accurate. Additionally, at a minimum, those using AI should know how to use it. In many cases, it is crucial for the researcher or clinician to understand the dataset, algorithm, and model. For example, when using surgical robotics, clinicians' knowledge about how to use the tool is paramount. Their knowledge about coding and how the engineers developed the tool may be slightly less important than their ability to use it, yet still certainly worth learning about. They especially should understand the AI tool's basics as well as its limitations.

In many cases, clinicians must understand and be able to convey to consumers the model's architecture, purpose, and training data source. While many suggest that clinicians will need to reskill significantly, AI is a broad category—some clinicians may not need to reskill while others may find that AI alters their job, replacing some tasks altogether. It may not be necessary for every clinician to do a deep dive into AI for health care at the expense of seeing patients and performing non-AI clinical tasks that healthcare consumers have come to expect. AI is impacting many aspects of care. Medical education is incorporating AI and will inevitably shift the field over time.

The framework is meant to organize and direct thinking about AI. It is not meant to lead to simple box checking. The levels of disclosure (1–3) are general to avoid creating blanket disclosure templates. Disclosures should be product specific when possible. Like informed consent to non-AI medical treatment, the framework aims to narrow the gap between the clinician's knowledge and the consumer's knowledge. Consumers generally do not need the full medical school experience to get the gist and the important details about their medical care. AI is similar—consumers can make medical decisions based on important points about the AI tool.

A. Zimmerman, *Informed Consent for AI and Health Care*,
SpringerBriefs in Ethics, https://doi.org/10.1007/978-3-032-11347-4_9

Likewise, clinicians should not all become engineers. If they were to, some of the science the public relies on would likely be compromised—the computers would become the scientists. AI touches on many aspects of health care, but it has not replaced typical clinical care. As it improves and more products become available and accessible, its use in health care is predicted to grow significantly. By developing a reasonable understanding of the tools they use, clinicians will be able to make material disclosures to consumers.

9.2 Significant Room for Improvement to Process and Forms

The use of lengthy forms for informed consent is widely criticized. Consent fatigue, fine print, and forms that equate important warnings to extraneous information undermine healthcare consumers. Verbal explanations of key information help with transparency. It would be great if IRB folks or clinicians drop the term "consent you" as if it is an action verb to be done to the healthcare consumer. People do not consent other people. Seeking informed consent allows the consumer to *consent to* the treatment.

In documents, important sections should be in boxes, bold type, and employ either bullet points or table formats. Data, billing, and treatment should be in separate documents ideally. Informed consent gets a bad rap partly because stealth small print contains important information. Furthermore, burying important warnings in long documents makes finding what really matters a search for a needle in a haystack. For clinical trials where a lot of information is necessary to the decision, bold bullet points, boxes, and other tools which break up monotonous pages help consumers. Additionally, initialing clear bold statements in documents is helpful. One lone signature at the bottom of a long document rarely indicates a consumer read every word. When talking to the healthcare consumer about the intervention, having various spots to sign or initial does more to ensure the main point of each section was noted and absorbed. Lastly, everything I aim for here concerns clear disclosures to promote voluntary informed decision making.

9.3 Conclusion

Informed consent may hold the key to unlocking AI's potential. In some cases, the ethical use of AI requires obtaining informed consent from the healthcare consumer. Healthcare consumers must have the information they need to decide whether and how to proceed.

A reasonable person standard helps clinicians avoid the subjective nature of trying to reveal or conceal information based on what each healthcare consumer would find material, be likely to understand, and want to know. Many people want as much information as possible while others want close to no information. Based on what a reasonable person would want to know when assessing risks and making medical decisions, the framework set forth here avoids the need for clinicians to assess individuals' level of education, ability to absorb the information provided, or risk appetite. Whether one is suspicious of AI or unquestioningly embraces it, clinicians and researchers should provide a baseline disclosure and obtain the consumer's consent as a standard procedure. In addition, prior to providing consent, consumers should have the opportunity to ask follow-up questions about AI tools. The informed consent process does not absolve clinicians of liability for using unvetted, unreliable, invalid, or otherwise flawed systems or for using well designed systems incorrectly or for tasks for which they were not designed. Requiring informed consent would not incentivize straying from medical standards.

In cases where informed consent is ethically compelled, an approach to the uncertainty about what to disclose requires a comprehensive yet easily accessible framework for evaluating AI tools. The framework presented here requires assessing the likelihood and degree of risk associated with an AI tool and the likelihood that clinicians overly rely on it. *This framework is unique in that it accounts for the high risk of overreliance on AI's output*. There is a noted push in the scant medical and bioethics literature to relax informed consent claiming AI is behind the scenes, too difficult to explain, or unimportant to the healthcare consumer. Yet reasonable people do want to know about the use of AI and may find a lack of disclosure deceptive. The framework proposed here is one way to respect healthcare consumers, keep them informed, and encourage dialogue.

Informed consent is not a deterrent to innovation. Healthcare consumers want more treatment choices, better drugs, and trustworthy hospital systems. Disclosures about AI provide an opportunity to showcase new developments. Considering the onslaught of new AI tools, some of which could be outcome determinative for healthcare consumers suffering from diseases, now is not the time to walk away from informed consent, a cornerstone of autonomy.

MIX
Papier aus verantwortungsvollen Quellen
Paper from responsible sources
FSC® C105338

If you have any concerns about our products,
you can contact us on
ProductSafety@springernature.com

In case Publisher is established outside the EU,
the EU authorized representative is:
Springer Nature Customer Service Center GmbH
Europaplatz 3, 69115 Heidelberg, Germany

Printed by Libri Plureos GmbH
in Hamburg, Germany